MAILING ADDRESS
Maritime Studies Research Unit
Memorial University of Newfoundland
St. John's, Newfoundland A1C 5S7, Canada

RESEARCH IN MARITIME HISTORY
NO. 51

CROSSING THE BAR: AN ORAL HISTORY OF THE BRITISH SHIPBUILDING, SHIP REPAIRING AND MARINE ENGINE-BUILDING INDUSTRIES IN THE AGE OF DECLINE, 1956-1990

Edited by
Anthony Slaven and Hugh Murphy

International Maritime Economic History Association

St. John's, Newfoundland
2013

ISSN 1188-3928
ISBN 978-1-927869-01-7

Research in Maritime History is available free of charge to members of the International Maritime Economic History Association. The price to others is US $25 per copy, plus US $5 postage and handling.

Back issues of *Research in Maritime History* are available:

No. 1 (1991)	David M. Williams and Andrew P. White (comps.), *A Select Bibliography of British and Irish University Theses about Maritime History, 1792-1990*
No. 2 (1992)	Lewis R. Fischer (ed.), *From Wheel House to Counting House: Essays in Maritime Business History in Honour of Professor Peter Neville Davies*
No. 3 (1992)	Lewis R. Fischer and Walter Minchinton (eds.), *People of the Northern Seas*
No. 4 (1993)	Simon Ville (ed.), *Shipbuilding in the United Kingdom in the Nineteenth Century: A Regional Approach*
No. 5 (1993)	Peter N. Davies (ed.), *The Diary of John Holt*
No. 6 (1994)	Simon P. Ville and David M. Williams (eds.), *Management, Finance and Industrial Relations in Maritime Industries: Essays in International Maritime and Business History*
No. 7 (1994)	Lewis R. Fischer (ed.), *The Market for Seamen in the Age of Sail*
No. 8 (1995)	Gordon Read and Michael Stammers (comps.), *Guide to the Records of Merseyside Maritime Museum, Volume 1*
No. 9 (1995)	Frank Broeze (ed.), *Maritime History at the Crossroads: A Critical Review of Recent Historiography*
No. 10 (1996)	Nancy Redmayne Ross (ed.), *The Diary of a Maritimer, 1816-1901: The Life and Times of Joseph Salter*
No. 11 (1997)	Faye Margaret Kert, *Prize and Prejudice: Privateering and Naval Prize in Atlantic Canada in the War of 1812*

No. 12 (1997) Malcolm Tull, *A Community Enterprise: The History of the Port of Fremantle, 1897 to 1997*

No. 13 (1997) Paul C. van Royen, Jaap R. Bruijn and Jan Lucassen (eds.), *"Those Emblems of Hell"? European Sailors and the Maritime Labour Market, 1570-1870*

No. 14 (1998) David J. Starkey and Gelina Harlaftis (eds.), *Global Markets: The Internationalization of The Sea Transport Industries Since 1850*

No. 15 (1998) Olaf Uwe Janzen (ed.), *Merchant Organization and Maritime Trade in the North Atlantic, 1660-1815*

No. 16 (1999) Lewis R. Fischer and Adrian Jarvis (eds.), *Harbours and Havens: Essays in Port History in Honour of Gordon Jackson*

No. 17 (1999) Dawn Littler, *Guide to the Records of Merseyside Maritime Museum, Volume 2*

No. 18 (2000) Lars U. Scholl (comp.), *Merchants and Mariners: Selected Maritime Writings of David M. Williams*

No. 19 (2000) Peter N. Davies, *The Trade Makers: Elder Dempster in West Africa, 1852-1972, 1973-1989*

No. 20 (2001) Anthony B. Dickinson and Chesley W. Sanger, *Norwegian Whaling in Newfoundland: The Aquaforte Station and the Ellefsen Family, 1902-1908*

No. 21 (2001) Poul Holm, Tim D. Smith and David J. Starkey (eds.), *The Exploited Seas: New Directions for Marine Environmental History*

No. 22 (2002) Gordon Boyce and Richard Gorski (eds.), *Resources and Infrastructures in the Maritime Economy, 1500-2000*

No. 23 (2002) Frank Broeze, *The Globalisation of the Oceans: Containerisation from the 1950s to the Present*

No. 24 (2003) Robin Craig, *British Tramp Shipping, 1750-1914*

No. 25 (2003) James Reveley, *Registering Interest: Waterfront Labour Relations in New Zealand, 1953 to 2000*

No. 26 (2003) Adrian Jarvis, *In Troubled Times: The Port of Liverpool, 1905-1938*

No. 27 (2004) Lars U. Scholl and Merja-Liisa Hinkkanen (comps.), *Sail and Steam: Selected Maritime Writings of Yrjö Kaukiainen*

No. 28 (2004) Gelina Harlaftis and Carmel Vassallo (eds.), *New Directions in Mediterranean Maritime History*

No. 29 (2005) Gordon Jackson, *The British Whaling Trade*

No. 30 (2005) Lewis Johnman and Hugh Murphy, *Scott Lithgow: Déjà vu All Over Again! The Rise and Fall of a Shipbuilding Company*

No. 31 (2006) David Gleicher, *The Rescue of the Third Class on the Titanic: A Revisionist History*

No. 32 (2006) Stig Tenold, *Tankers in Trouble: Norwegian Shipping and the Crisis of the 1970s and 1980s*

No. 33 (2007) Torsten Feys, Lewis R. Fischer, Stéphane Hoste and Stephan Vanfraechem (eds.), *Maritime Transport and Migration: The Connections between Maritime and Migration Networks*

No. 34 (2007) A.B. Dickinson, *Seal Fisheries of the Falkland Islands and Dependencies: An Historical Review*

No. 35 (2007) Tapio Bergholm, Lewis R. Fischer and M. Elisabetta Tonizzi (eds.), *Making Global and Local Connections: Historical Perspectives on Ports*

No. 36 (2008) Mark C. Hunter, *Policing the Seas: Anglo-American Relations and the Equatorial Atlantic, 1819-1865*

No. 37 (2008) Lewis R. Fischer and Even Lange (eds.), *International Merchant Shipping in the Nineteenth and Twentieth Centuries: The Comparative Dimension*

No. 38 (2008) Adrian Jarvis and Robert Lee (eds.), *Trade, Migration and Urban Networks in Port Cities, c.1640-1940*

No. 39 (2009) Henry T. Chen, *Taiwanese Distant-Water Fisheries in Southeast Asia, 1936-1977*

No. 40 (2009) John Armstrong, *The Vital Spark: The British Coastal Trade, 1700-1930*

No. 41 (2009) Carina E. Ray and Jeremy Rich (eds.), *Navigating African Maritime History*

No. 42 (2010) S.G. Sturmey, *British Shipping and World Competition*

No. 43 (2010) Maria Fusaro and Amélia Polónia (eds.), *Maritime History as Global History*

No. 44 (2010) Silvia Marzagalli, James R. Sofka and John J. McCusker (eds.), *Rough Waters: American Involvement with the Mediterranean in the Eighteenth and Nineteenth Centures*

No. 45 (2011) Jaap R. Bruijn, *The Dutch Navy of the Seventeenth and Eighteenth Centuries*

No. 46 (2011) Lewis R. Fischer and Even Lange (eds.), *New Directions in Norwegian Maritime History*

No. 47 (2011) John Armstrong and David M. Williams, *The Impact of Technological Change: The Early Steamship in Britain*

No. 48 (2012) Ralph Davis, *The Rise of the English Shipping Industry in the Seventeenth and Eighteenth Centuries*

No. 49 (2012) Gordon Boyce, *The Growth and Dissolution of a Large-Scale Business Enterprise: The Furness Interest 1892-1919*

No. 50 (2013) Torsten Feys, *The Battle for the Migrants: The Introduction of Steamshipping on the North Atlantic and Its Impact on the European Exodus*

Research in Maritime History would like to thank Memorial University of Newfoundland for its generous financial assistance in support of this volume.

Table of Contents

Interviews

Lower Clyde

Upper Clyde

The Tyne

The Wear

Barrow-in-Furness

The South Coast

The Humber

Belfast

British Shipbuilding Industry Officials

The Trade Unions

The Civil Servants, Board of Trade, Shipbuilding Enquiry Committee, Shipbuilding Industry Board, Ministry of Technology, Department of Trade and Industry, Department of Industry

Politicians

British Shipbuilders Plc

About the Authors

ANTHONY SLAVEN <Tony.Slaven@glasgow.ac.uk> is Emeritus Professor of Business History and former Director of the Centre for Business History in Scotland at the University of Glasgow. He is Past-President of the Institution of Engineers and Shipbuilders in Scotland (1992-1994); Past-President of the European Business History Association (1995-1997) and currently Director of the Ballast Trust. His publications include *British Shipbuilding 1500-2010: A History* (Lancaster, 2013); *Reviews of United Kingdom Statistical Sources Vol. XVI: Iron and Steel Shipbuilding* (Oxford, 1984, with David W. Heal); *Dictionary of Scottish Business Biography* (2 vols., Aberdeen, 1986-1990, edited with S.G. Checkland); *From Ledger Book to Laser Beam: A History of the Trustee Savings Bank in Scotland from 1810 to 1990* (Edinburgh, 1992, with Michael S. Moss); and *The Development of the West of Scotland, 1750-1960* (London, 1975; New ed., London, 2009). He is also the author of many articles on shipbuilding and business history in general.

HUGH MURPHY <hmurphy33@hotmail.com> is Professor of Maritime Business History in the Centre for Business History in Scotland at the University of Glasgow. He was Hon. Editor of *Mariner's Mirror*, the international quarterly journal of the Society of Nautical Research, from November 2005 to November 2012, and is currently Visiting Reader in Maritime History at the National Maritime Museum, Greenwich. He is the author or co-author of many books and articles on the British shipbuilding industry and business history generally, including *British Shipbuilding and the State since 1918: A Political Economy of Decline* (Exeter, 2002, with Lewis Johnman), which won the Anderson Medal for Research in Maritime History for 2002. Professor Murphy has held many research fellowships, including the Caird Junior and Senior Research Fellowships at the National Maritime Museum, Greenwich.

For Our Shipbuilders and all Who Worked with Them

Preface
A Shipbuilding Libretto

Since 1959, the Colquhoun Lecturer in Business History at Glasgow University, initially Peter Payne and subsequently Tony Slaven, had been searching for, and saving from destruction the records of Scottish industry, including the commercial and technical records of shipbuilding firms. This enterprise accelerated in the 1960s and 1970s as the decline of traditional heavy industries gained momentum. Consequently, we had a rich and varied collection of shipbuilding company records but little information on the men who owned, managed and operated the firms. The establishment and funding of the Centre for Business History in Scotland in 1988, and my appointment as Director and Professor of Business History, created the framework and resources to undertake the research focussing on the recollections of major players in the history and development of British shipbuilding since the Second World War.

Ironically, it was the winding up of the nationalized British Shipbuilders Plc that opened up access to individuals involved in all the shipbuilding districts and those who had been instrumental in the creation and management of that state corporation. Dr. (now Professor) Ian Buxton (Denny Brothers, Dumbarton and later University of Newcastle) had been collaborating with Tony Slaven and Fred Walker (Denny Brothers, Fairfield, Govan, Connell, Scotstoun, Hall Russell, Aberdeen, and National Maritime Museum, Greenwich) in an attempt to collect information on the history of British shipbuilding, and he persuaded British Shipbuilders Plc to provide seed funding that enabled two initiatives to get underway. Ian Buxton's project was to produce a database of all British-built ships since the nineteenth century, while mine was to record the recollections and opinions of men involved in all aspects of the industry since 1945. Professor Buxton's "British Shipbuilding Database" is now complete and is a uniquely detailed record of 80,000 British-built ships in the nineteenth and twentieth centuries. It is now available for consultation in the Marine Technology Special Collection in Newcastle University's School of Marine Science and Technology and was formally opened on 23 May 2012.

The oral history project was designed and based in the Centre for Business History in Scotland. In the final days of British Shipbuilders we were acutely aware that if we did not move swiftly to capture a libretto from those still involved and those who had retired, the life-long experiences of these men would be lost forever. Designing a wide-ranging questionnaire was not difficult for an industry which had been subject to decades of public scrutiny and had been under public ownership and management in the last years of its existence. The men selected were in part dictated by who was still active and in large measure by recommendations from within the firms and districts.

The resources of the Centre for Business History, and the generosity of British Shipbuilders Plc, made it possible for me to appoint a Research Fellow on a three-year contract to assist in designing the project and to undertake the interviews. The researcher appointed was Philip Taylor, now Professor of Human Resource Management at Strathclyde University. Months of designing the questionnaire and writing to potential interviewees produced a list of sixty men, many of whom had careers in more than one company or had worked in two or more of the main shipbuilding districts. Many initial interviews were undertaken by myself and Philip Taylor as a team. This allowed us to refine the questionnaire to encourage spontaneity and be confident that we had a workable framework which provided structure while encouraging the contributor to range widely over the issues. Once this was settled, the main task of interviewing fell to Philip Taylor, who travelled the country to talk with our selected interviewees, all of whom had seen the questionnaire in advance and had agreed to have the discussion recorded. We then had the tapes transcribed and the text returned to the contributor for correction and editing. All the recordings were made in 1991 and 1992, providing the large volume of transcripts from which the versions here have been edited and organized.

By 1993 we had collected a massive amount of information in interview texts of substantial length. Interviews could last up to two hours. Given the volume of the transcripts and the currency of the commentaries, I decided that it was appropriate to let the data rest before further editing and publication. As an historian, I was conscious of the need to stand back from events to be able to take a more objective view.

The challenge of editing and organizing the interviews for publication only began with the appointment of Hugh Murphy as a Research Fellow in the Centre for Business History in Scotland on a part-time basis. Hugh had many responsibilities but was immediately attracted by the challenge of giving a structure to the large body of oral evidence. This has been a protracted exercise, largely owing to my other commitments. The final editing and organization of these unique librettos has been carried out mainly by Hugh Murphy, without whom they would never have been brought to publication. While I was the originator, guide and mentor throughout the project, and am responsible for its content, it owes its final appearance to the energy and commitment of Philip Taylor at the outset, and to the skill and perseverance of Hugh Murphy in bringing it to a conclusion. I am deeply indebted to them both.

Anthony Slaven
Emeritus Professor of Business History
University of Glasgow
July 2012

Introduction

This book contains an edited collection of sixty interviews with the major players in the British shipbuilding, ship repair and marine engine-building industries, as well as those in government and the civil service responsible for this sector, from the 1960s to the 1980s. The tale that unfolds in their own words is set against a background of the long-term, relative decline of an industry that easily led the world in output from before the First World War until it was successfully challenged, and then surpassed, by international competitors from the mid-1950s onwards. The participants range from the managing directors of family-owned firms and public companies to national trade union leaders, senior politicians and civil servants. All witnessed or were intimately involved in a hitherto unparalleled downturn in the industry's international competitiveness which eventually resulted in the end of volume shipbuilding in the United Kingdom before 1990.

The interviews were undertaken in 1991 and 1992. They were informal, with no time limits. Indeed, some of the transcripts stretched to over forty pages, with the average being twenty-two. Questions, *inter alia,* covered the interviewees' experience of the industry, the firm, marketing, industrial relations, relations with shipowners, modernization and investment, attitudes to international competition, views on the role of the state and the government's response to the industry under pressure.

Of the sixty participants, forty-three were directly involved with shipbuilding, marine engineering or ship repair. Of these, eleven were directly involved with state-owned British Shipbuilders Plc from 1977 onwards. Five were national politicians, three were trade unionists, four were senior civil servants, three were members of shipbuilding organizations, one was Chairman of the Shipbuilding Enquiry Committee of 1965-1966 and one was a civil engineer involved in shipbuilding projects. All participants had wide-ranging educational and occupational backgrounds. The largest professional group interviewed were naval architects (thirteen). Nineteen participants rose to become managing directors. Seven participants were chairmen, of whom three chaired British Shipbuilders. Eight participants had engineering backgrounds. Five began as ship draughtsmen. Three became finance directors, and two were industrial relations/personnel directors. Of all the participants only three – Sandy Stephen (Alexander Stephen, Linthouse), Harold Towers (Readhead, Tyne) and Paddy Christie (Swan Hunter and Wigham Richardson, Tyne) – had family connections, and all for varying periods became managing directors. Four participants served apprenticeships, and one of these, Bob Easton, rose to become a managing director.

We have information on the education of nearly half the respondents. Of these, three had a high school education, and two completed grammar school. Another two gained qualifications from technical colleges. At the university level, seven took degrees at Glasgow University, six from Cambridge, three from Oxford, two from each of London and Newcastle, and one each from Durham and the University of Aberdeen. The educational background of the others is unknown.

In this project we decided to approach the experience of the industry from the perspective of the men who led and managed it. The hope was to gain an insight into their understanding of the economic and political circumstances, and of the competitive environment in which they worked. These men were drawn from the major shipbuilding, ship repair and marine engineering firms; this was designed to span the spectrum of experience in the industry. The industry focus was then complemented by approaching the same questions and responses for interviews with people drawn from the civil service, ministers of state and officials in the industry trade associations and the trade unions.

A few of the men interviewed had management experience in the industry from the 1930s, but most came into management in the 1940s and 1950s. Of these, many had retired, and some have since "crossed the bar" and died. Others, forming a new management cadre, had come into senior positions from the 1970s. Not surprisingly, there was some difference of opinion across the generations, but the reader will note that in spite of that there is a broad consensus about the problems and responses of the industry as it faced increasing competition and lost out to newer rivals.

Given the location of most of the larger firms in the UK shipbuilding industry in three major river centres – the Clyde on the west coast of Scotland, and the Tyne and Wear on the North East coast of England – the majority of interviews were undertaken in these areas. In addition, we also interviewed senior personnel in the two major firms on the North West coast of England, Cammell Laird at Birkenhead and Vickers at Barrow-in-Furness, as well as Harland and Wolff at Belfast, and also the only naval firm on the South coast, Vosper Thornycroft. We also interviewed the managing director of a small shipbuilding firm on the Humber.

The initial approach in the interviews was to invite the men to outline their background, how they had come into or been associated with the industry, and how their career paths had developed. Underpinning the subsequent answers to questions was a sense of *déjà vu* regarding international competition, and from this, interrelated topics such as labour difficulties, modernization, marketing and shipowners' attitudes naturally flowed. Despite this, however, with the benefit of hindsight a minority of those interviewed advanced from the outset a fundamental critique of the UK shipbuilding industry's deep-seated failings. In their view, the worship of tradition bedevilled the industry, a tradition that cemented arrogance even in the face of overwhelming evidence that other nations were in fact building ships better, cheaper and faster than was the case in the UK.

Methods of ship production in the UK were, certainly in contrast to Japan, Sweden and later, South Korea, antiquated. UK yards eschewed increases in capacity, lacked capital investment and were wary of the expense of constructing building docks; they continued to build on slipways and were very slow to adopt modern fabrication methods. There were no production control systems worthy of the name, and no production engineering; basically, the sole measure of output and progress was the ship on the berth. By the time concepts such as flow-line production and critical-path analysis were being discussed, other nations, particularly Japan and Sweden, had been proficient in them for many years. Moreover, some felt that the small scale of individual firms in the UK disadvantaged them in relation to larger, vertically integrated continental and East Asian firms.

Nonetheless, others interviewed felt that being bigger was not necessarily better, and that remaining small was a distinct advantage in that management had clear lines of communication and could react quicker to changes in demand. Still, approximately eighty percent of those interviewed regarded the lack of capital investment in UK shipbuilding as a problem. Some thought this to be a central weakness, while others thought it less decisive. UK yards on the whole had historically been less capital intensive than their continental competitors, and modernization after 1945 in spatially constrained yards had been held up by full order books stretching in some cases ten years ahead as world trade exploded during the long postwar boom. While capacity increased elsewhere, UK shipbuilding stood still; the industry basically rested on its laurels and on the whole missed the boat on the growth areas of world shipbuilding: large tankers and bulk carriers.

By 1965, however, some world-famous firms, such as Denny at Dumbarton and William Gray at West Hartlepool, had gone into liquidation, and the government, which had not intervened in the industry but did closely monitor it, created a Shipbuilding Inquiry Committee aimed at enhancing international competitiveness. At that stage the industry employed approximately 140,000 people. The resulting report, colloquially named after its chairman, Reay Geddes, was published in March 1966. The Committee's main recommendations were the large-scale grouping of firms by specialization on river centres and the rationalization and separation of marine engine-building from shipbuilding. Geddes proposed that this could be achieved by grants and loans dispensed from a new government-sponsored Shipbuilding Industry Board. Although most of the respondents agreed that the Geddes Report contained some good ideas, somewhat predictably the warship-building sector was against grouping, but by this stage they were almost entirely wards of the state. On the mercantile side, the writing was on the wall in the sense that British shipowners were beginning to desert the domestic industry in increasing numbers. Nonetheless, the vast majority of respondents did not apportion blame on British shipowners but saw that shipping was truly an international business and that owners were entitled to look for a better deal on price, delivery and credit elsewhere.

These admissions, given against a backdrop where all firms previously had long associations with UK shipping lines – and basically built to order for them – was at least honest, if somewhat disingenuous. Marketing in UK shipbuilding, with some later exceptions, was rudimentary. Contracts were often internally agreed before being fully specified, and bespoke relationships with shipowners comprised a disproportionate amount of most companies' output. A prevailing attitude was that as long as world population continued to increase, demand for ships would remain at least constant. In consequence, a number of firms sat back and waited for orders rather than going all out to get them. One respondent commented that inadequate marketing was a polite way of saying that there was no marketing at all. Marketing in the sense of establishing a research team to look at future trends in ships and looking at a firm's strengths and weakness was virtually non-existent. It was left to chairmen or chief executives of individual firms to go out and try to sell their products, as it was felt that no one else was competent to do so.

What did take up a disproportionate amount of senior management's time was the resolution of trade disputes in the shipyards. These were many and varied. Hardly any yards had personnel managers, and in those that did, these managers had a very low status. Trade union representatives would not deign to talk to them and insisted on talking directly to higher management. In one yard, Scott's of Greenock, the managing director, the late George Hilton, who had died before this project was implemented, recalled that disputes over which trade had the sole right to undertake a particular process were legion, (demarcation) and that he would have not been surprised if the yard's sheet metal workers had claimed the exclusive right to open tins of soup in the works canteen!

The demarcation of trades had a long historical pedigree in the UK shipbuilding industry, and trade advantages were jealously guarded. As well as the class divide between management and men, inherent sectionalism within the workforce was a *de facto* element of life in UK shipyards. Indeed, some respondents predictably put the entire blame for the industry's demise squarely on the shoulders of the trade unions, while the vast majority more convincingly pointed to a combination of management and labour failings. One shipyard manager in his first five years at an Upper Clyde shipyard estimated that he spent some twenty-eight hours per week arguing with union representatives when he should have been exclusively involved in running the yard. Moreover, the multiplicity of trade unions was seen as a major problem in that the introduction of agreements on issues such as flexibility, interchangeability and other changes in work practices in a generally restrictive environment, were unnecessarily protracted in the late drive to modernize the industry.

Only from 1984 onwards, when the industry was on its knees and in danger of total collapse, did significant changes in working practices occur, by which stage it was arguably too late in any event. At this point, due almost entirely to nationalization from July 1977 onwards, the number of graduates in the industry had increased from a historically low base. Middle- and lower-management attitudes tended to reflect their backgrounds, which were mostly from within the industry. Authority without any real responsibility was the norm, and graduates from outside were viewed with suspicion in most yards where the cult of the practical man persisted. Shipyard foremen and middle managers tended to be hard-bitten individuals and captives of their own bitter experiences. Younger individuals, often frustrated by the lack of career advancement in yards with pared management structures, often left for new pastures, and in consequence, those that remained tended to reinforce the old rather than look to new methods. Up to nationalization in 1977, some degree of nepotism still existed, particularly in family-owned and controlled firms, one of which, Scott's of Greenock, then the world's senior shipyard, had a seventh-generation family member in control, as did Stephen of Linthouse. Nationalization was viewed by some to be inevitable and by others as unnecessary, as was privatization from 1984 onwards.

These then, are just some of the interesting topics covered in the sixty interviews. Just how accurately they reflect the problems and successes of the UK shipbuilding industry over three decades is of course open to the reader's interpretation. We set out to try to put into context the decline of the industry through the systematic canvassing of views and opinions of those intimately involved at the higher levels within it and without. Analysis of the complimentary interviews in the latter part of this study with senior civil servants and politicians has revealed differing perspectives and conflicting viewpoints, and readers are encouraged to compare them with their own perspectives and understanding of the industry.

What is beyond doubt is that, despite numerous personal caveats, hindsight, loss of memory over time and the exaggerations and the ingrained prejudices of individuals colouring the evidential value of their contributions, we have in our possession a unique historical archive of opinion which we now set out in summary first-person form to enable further interpretation and dissemination to the wider historical community and the world at large. When a particular point in the text is unclear or demands clarification or explanation, we have commented in italicized parenthesis to aid the reader's contextual grasp.

Interviews
Lower Clyde

Interview 1: Alexander Ross Belch, Managing Director, Scott Lithgow Ltd., Greenock and Port Glasgow, and Bowling, Dunbartonshire

Scott Lithgow Limited was created by February 1970 as a result of a merger of the Greenock and Port Glasgow shipbuilding and engineering interests of Scotts of Greenock (est. 1711) and Lithgows Limited of Port Glasgow (est. 1918). Ross Belch's father, Alexander Ross Belch Sr., was a leading light in the shipbuilders' commercial organization, the Shipbuilding Conference (est. 1928).

I began my career in shipbuilding by way of an apprenticeship at Fairfield at Govan in 1938, and combined this with a degree course in Naval Architecture at the University of Glasgow during the war years. In1945, I became an assistant manager at Fairfield, before shortly moving downriver to Lithgows Limited, where I was subsequently appointed Director and General Manager in 1954. By 1967, both Scotts and Lithgows had announced their intention to merge, and by the time of the merger in 1970, I was in charge. I led the merged company through to nationalisation under British Shipbuilders plc in July 1977; and stayed on as managing director and a director on the Board of British Shipbuilders, but subsequently retired in 1980.

The 1940s and 1950s were pretty easy days in shipbuilding; German and Japanese yards were not allowed to build ships after the war. Many of the European yards had been bombed. It was not difficult to get work especially as Lithgows were a bread and butter yard, and completed virtually every contract at a fixed price. This led to tight cost controls, controls that were, by and large, absent in cost plus mixed naval and mercantile yards. Lithgows ran a leaner and tighter ship.

On the whole, labour and management conservatism cancelled each other out. I carried out labour negotiations as well as running the yard. In the post-war period you could make a deal with national officials, but there was no guarantee that it would stick at local level. Labour relations were very much my life as well as trying to sell ships, produce them, and design them. You had to know the whole game if you were to be a reasonable managing director...I think it was inbred, the craft system; the demarcations had grown up in times prior to the war. To an extent even though you could convince people that they would not lose their jobs, it did not matter. It was bloody-mindedness, a "thrawness" we have on Clydeside. The Boilermakers motto was "we have what we hold."

The whole shipbuilding industry was pretty old, and we had all these guild and craft systems. We had a situation whether we liked it or not. It had been a tough time in the 1920s and 1930s, which I did not know, but it was hire and fire. There were attitudes inbred, and it took a long time to try and break them down. I always like to think that we did a lot to do that at Lithgows. I like to think in many ways we led the way.

When I was Managing Director of Scott Lithgow, in charge of six shipyards, I used to meet all the union shop stewards separately. It probably was divide and rule, and they knew it, but it took a long time for them to get together. One of the basic problems with the Clyde, and I think I am being objective, was that the Upper Clyde yards, generally, for whatever reason, had experienced life too easily after the war. They gave way too much on wage demands. They paid out too readily. They would not face sometimes the necessity of a strike in those days. They were weighing up the costs of a strike and the penalties for delay [*of the ship*] and the damage done to reputation. They would always give way to the cranemen who were holding them to ransom.

The result was that although we had a Clyde Shipbuilders Association, which considered labour disputes, we were always fighting on the Lower Clyde. We found that time and time again, Upper Clyde yards were paying out money that they could not afford to pay. We eventually formed our own association, the Lower Reaches Employers Association. This was one of the big troubles that we faced. Understandably, our boys were looking for [*pay*] parity with those on the Upper Clyde. There was an unwritten agreement that nobody gave a penny an hour extra to his welders. We always held together. The wage rates were reasonable; they were not starvation rates, but they were rates that we could afford to pay and stay alive to get work. We watched the rates soaring away on the Upper Clyde, until 1971, they were paying £8 a week more to skilled labour. We had a strike of Boilermakers on the Lower Clyde on the simple issue of parity, and I said no as we could not afford to give it, and as a result the Boilermakers were on strike for seven weeks. Danny McGarvey [*the leader of the Boilermakers Society*] came up from Newcastle and we made a small gesture which was peanuts. He had foolishly supported it as an official strike, and his money was running out. He knew it, and I knew it, and he got peanuts from me just to save his face. He put it to the men, and they rejected it. They met again on Monday, and he told them that they had to accept it, and they did.

There was a considerable lack of graduates. I was an oddity coming down as a graduate. People were not really properly trained as they would be today. A lot of them were good practical shipbuilders who had come up from the grass roots. I think many of them were very good. They were like sergeants in the Army. When they got promoted they sat on the boys under them. People say that [*the industry*] was reactionary and stupid. That was proved not to be the case many times when the Government introduced into our industry whiz kids who made a greater mess than any of our practical shipbuilders.

By 1959, [*by which stage the long post-war sellers market in shipbuilding had been replaced by a buyers market*] the Germans were on the go, the Japanese were about to get going, [*Japan led the world in shipbuilding output for the first time in 1956, displacing Great Britain for good as the world's greatest shipbuilder*] and the Continent was on the go with fairly updated facilities. In 1959, Jebsen came in with an order for two 15,000 ton bulk carriers. We got the order against massive competition by cutting our price. I was over in Bergen negotiating for two or three days, and we took a cut on the price. We had this design, which other people had too, of the top wing tanks and the bottom hopper tanks, which ultimately became the traditional bulk carrier design. We made that contract. It was £950,000 and we lost £50,000 on each ship. We kept the yards going. We just had to keep hunting, and we did quite a bit of designing ourselves. This is what I call "true marketing," telling people that this was the ship they wanted, as opposed to waiting until they came to you to tell you what they wanted.

By 1962, the Government was getting worried about the problems of the shipbuilding industry, and appointed a lady, [*Lord Piercy's daughter*] in some government institution to see what was wrong. She produced a report which very much upset the shipbuilders, [*Report of Department of Industrial and Scientific Research Report on the Research and Development Requirements of the Shipbuilding Industry, 1960*]. It was about poor management, family dynasties. The shipbuilders were furious about it and appointed a committee of young shipbuilders supposedly under the chairmanship of a man who had just retired from the ship repairing industry, called James Patten, to look into the industry ourselves. It was an industry study of the industry. I was one of the people on the committee, and so was Derek Kimber. He was the yard manager at Fairfield, when I was down river. Some of the Patten Report [*Productivity in Shipbuilding (London, 1961)*] is as good today as it was when it was written in 1961. We toured continental yards [*four large German yards; two large, one medium and six small Dutch yards; one large Danish yard; and four large and one medium Swedish yards*] to see what others were doing.

I should say that we were not standing still on the Lower Clyde. We were developing the yards. We were spending more money, particularly on our new fabrication facilities. We were maybe slow to move to welding, but we did move slowly but surely. I was perhaps a bit conservative myself. I did not like to change too quickly because I thought you could only change at the pace people who had to accept the changes would take. That is an automatic thing in everybody, and I am guilty of it myself that when somebody asks you to do something new you immediately look for the reasons why you can not do it. We did, though, do a lot of modernisation in those early days at Lithgows.

On modernisation we were as far ahead as anybody. Stephen's yard [*Linthouse, Upper Clyde*] was pretty good. I think they overdid it. John Stephen was something of an idealist. But they did modernise and so did Connell's [*Scotstoun, Upper Clyde*]. John Brown's [*Clydebank, Upper Clyde*] did not, frankly; and Fairfield [*Govan, Upper Clyde, owned by Lithgows*] did not do much. Yarrow [*Scotstoun*] was doing something, but they were essentially on the naval side. Other yards like Blythswood [*Scotstoun*] went out of business. Harland and Wolff at Govan closed. That was the period when yards were going out when the storm hit them in the 1960s.

I was Managing Director of Lithgows when I was forty-three, and two years later the Geddes Report was published. A lot of Geddes was good. It was the way it was carried out afterwards. Geddes, for example, suggested that we [*the shipbuilding industry*] were somewhat not integrated, that the yards had grown up topsy-turvy. You can not criticise anyone for that because they were old yards. Many of them needed modernising. The 1960s was the decade of "get big and you are efficient" idea, which was rubbish. I think he was right in suggesting certain mergers.

They made a mess of the implementation because they tried to mix up oranges and lemons and say that they had a common fruit. We saw it happening on the Clyde. Again my comment is "that Glasgow is the Clyde and the Clyde is Glasgow." I have seen that attitude living in Greenock [*twenty-five miles west of Glasgow*]. When Geddes came through first with all his merging proposals, it made sense providing sensible arrangements for merging we made. I do not think that they were. It made sense that they [*the Shipbuilding Industry Board*] made money available, which we at Lithgows took a large chunk of for further modernisation. Like when we built that big crane [*actually built by Sir William Arroll and Son, delivered very late, and used in Scott Lithgow's Glen yard*] and the berth [*actually a large sloping concrete mat*] to build big tankers [*in half sections*] and so on.

Somebody had the idea that Geddes said that there should be a single group of shipbuilding yards on the Clyde. That was not what Geddes said. I remember exactly what he said, and I had it word perfect in those days. They were words to the effect that there should be no more than two groups. It [*the Geddes Report*] certainly did not say that there should be a single group. The Upper Clyde yards messed around. We saw immediately that our neighbours, Scott's, were our natural partners. Ours was a happy merger, which worked extremely well. We were lucky that Mr. Hilton [*George Hilton, Scott's Managing Director*] was in his sixties and I was in my forties. The two yards were very complimentary. Scott's did the naval building and the more sophisticated tonnage, and Lithgows the bread and butter mercantile tonnage. By this time whilst Geddes was deliberating, Scott's had taken over the neighbouring

Greenock Dockyard and we [*Lithgows*] and taken over Ferguson Brothers at Newark, Port Glasgow.

The problem basically with British shipbuilding was that as Geddes said the smallness of the yards, and the fact that the capital base was not large. I am not critical of the families, but many of them tended to be family businesses, and were not very modern. Suddenly on to the scene came particularly the Japanese with new yards, new techniques and everything on a massive scale. The competition from Japan was such that although you could keep your shipyards going, you would never make much money out of it. It was never an easy business. Going through the 1960s was a very tough time to keep your yards going at almost break even prices. That was the reason that I personally made the decision to switch in the early 1970s to something else in our yards other than building ships. Offshore work was developing from the North Sea, and we made a conscious effort to move into the offshore world, with, first, building two offshore supply boats at Ferguson Brothers for Seaforth Marine, and at the same time we were negotiating with the Ben Line, from which ultimately came an order for a dynamically positioned drill ship [*Ben Ocean Lancer*]. And then the Norwegian shipowner, Atle Jebsen came along and ordered another [*Pacnorse 1*]. Then the last contract I did before I pulled out [retired] in 1980 was the first semi submersible emergency support vessel [*Iolair*].

I do not think that competition [*from abroad*] was unfair. I do not make the excuse that the competition is always subsidised. It was very difficult to tell with Japan. Japan always had that expression, "Japan Incorporated." The government, the steelmakers and the shipbuilders would do it together. There might be special hidden things, for example, no one outside of Japan built ships for the Japanese. I did not shelter behind the excuse that they benefited from hidden subsidies. They were highly efficient, but I do not think that our own government could have supported our industry better. It became a whipping boy...there was always interminable investigation. Booz Allen (*British Shipbuilding in 1972: Report to the Department of Trade and Industry by Booz Allen and Hamilton, 1973*] probed around, but did not really get down to the nitty gritty. They were always saying we were a useless, inefficient lot. Booz Allen, [*nicknamed "the Yellow Peril" by shipbuilders by the colour of its binding*] I thought, was a load of tripe, and I told them that. I was by this time Chairman of the Shipbuilders and Repairers National Association.

There was an awful lot of theory in Booz Allen. I always remember them telling me that marketing was the thing. "Your Cartsdyke [*Greenock Dockyard*] yard, Mr. Belch: What ships do you build best there?" I told them that probably we built refrigerated cargo vessels. And they would say, "that's what you should build there." And I replied, "what if no one wants them?" It was theoretical soapflakes. Shipbuilding is a peculiar industry.

Prior to this Upper Clyde Shipbuilders [*a merger of John Brown, Connell, Fairfield, Stephen and Yarrow*] had gone bust. Those were the days of Nicholas Ridley and John Davies [*Conservative Ministers*], and this [*the shipbuilding industry*] was the lame duck of all lame ducks. That is why Maggie [*Margaret Thatcher, later Prime Minister in 1979*] never liked the industry, because that was [*the saving of a bankrupt Upper Clyde Shipbuilders by the Conservative government after a massive campaign from the workforce*] the first massive u-turn of any Tory Government. I thought that at the time that [*UCS*] should be killed, because they were causing such problems for us on the Lower Clyde. Michael Scott and I went to see the Government and civil servants and told them that UCS should not be further propped up, but civil servants believed that there would be "blood on the gutter if they did not."

In 1974, I became President of the Shipbuilders and Repairers National Association and it was my job to fight the nationalisation of the industry, which the Labour Government had proposed in that year. I made stacks of speeches saying what a bloody disaster it all was. I did not believe in it, and I was right. On the question that we could have carried on independently in the 1970s and 1980s, I do not know. Nationalisation is an inefficient method of operation. We all thought that, and knew that. That is why the Tories [*Conservative Government*] tried to de-nationalise [*from 1983 onwards*]. There are certain national utilities that should remain so. But, for an industry like shipbuilding to be nationalised is an inefficient way of working. Finally, after many hiccups, the second reading of the Bill to nationalise the industry went through, and after that we expected it to be inevitable. Graham Day [*a Canadian, and former Managing Director of Cammell Laird of Birkenhead*] was appointed Chief Executive Designate, and he said to me when I joined the Organising Committee; "the fight against nationalisation is over. The second reading is through!" My thought, stupidly now as I realise, was that you could not change the inevitability of it, but maybe I could influence the way it was going to be created along the lines of private enterprise. However, it was not to be. We soldiered through the hybridity [*if a Bill is deemed hybrid, then it fails in Parliament*] of 1976. I did my best to persuade Graham Day to stay, but he would not. For a time they [*the government*] had no Designate. I have to say they appointed a civil servant, Michael Casey, a quite wrong appointment as Chief Executive. I carried on with the Organising Committee for a period until they formed a Board, and I became one of the Board Members. That was the first Board [*in July 1977, when the industry was finally nationalised and was known as British Shipbuilders, with its headquarters in London*]. I got very disenchanted because it was going to be structured like the Civil Service. The papers started flowing, and the forms started to be filled. My mail doubled. Nobody took any responsibility. I decided to pack it in. It became a bureaucracy. It was pipedream to think that I could change it.

Interview 2: Johnston Robb, Engineering Director, Scott Lithgow Limited

I joined Scott's, which was an independent private company, in 1935, as an office boy, and served two years. At the age of sixteen years and three months I started an apprenticeship as a fitter. In parallel with that, I also attended evening classes. After two years and ten months of my apprenticeship in various departments, I was asked to go to the Drawing Office as a result of attaining a certain standard in evening classes. In August 1940, I became a trainee Draughtsman. In May 1941, during the Blitz, the whole place [the Drawing Office and Company Office] was burnt down. Fortunately the entire daily workload was transferred to concrete safes on a daily basis, and we were able to carry on. By 1948, I was an Assistant Engineering Estimator, and by 1950, a Personal Assistant to the Managing Director. In 1953, I was appointed Assistant Chief Draughtsman, and in 1955, Chief Draughtsman (Engineering). In 1960 I moved into line management as Engineering Manager of Scott's, and in 1966, became Engineering Director, and continued in that capacity post the Scott Lithgow merger in 1970.

I think that the major strength of Scott's was that it was a private company. The major shareholder was John Swire and Sons of London who owned the airline Cathay Pacific [*and the Taikoo Dockyard in Hong Kong*] amongst many other things. They tended to retain a very distinct influence on what the company did, and they certainly watched the financial side of the business. The second strength we had was our reputation for building high-class products, particularly submarines. The company was technically strong, and the drawing offices were particularly good. We were not that large, and I am not sure whether that was a strength or a weakness. The Scott part of Scott Lithgow never got above 3500 people. You have to understand that a great influence on all that went on in Greenock in the years from 1960 onwards was the navy. People like Yarrows and Vickers (*Barrow*) who always gave jobs to retired naval officers, seemed to have a far better image of the navy than we had. I felt this was a great shame because we did not deserve to be in that position. We were not strong in dealing with the navy and that was a weakness.

On the naval side, we saw ourselves as being in competition with Vickers and Yarrows, though in the latter case we were not really in competition because we got the submarines and Yarrow got the destroyers [*and later, frigates*]. It is difficult to say who our competitors were on the merchant side because we were on the edge of the bulk carrier boom. There was a tremendous preoccupation with bulk carriers and tankers. This had a marked effect on a lot of shipbuilders, particularly the smaller ones who could not build the large container vessels, each of which replaced five or six conventional cargo ships. We did, latterly, but in the medium size range.

There were a number of constraints on what we could do at Scott's. One was the number and length of berths. By 1960, we were committed to building two new large berths and new craneage, which were state of the art at the time. By 1960, we had done what seemed to bring us up into a competitive position for the medium to medium-large size of ship. For the Royal Navy we specialised in submarines and one-off highly technical ships, special purpose ships. We also built fleet replenishment ships. [*Royal Fleet Auxiliary*]

I think we realised that ships were going to get much bigger, and, coupled with the tremendous advance in prefabrication techniques; we felt that we had to do that. On the question of the small scale of firms highlighted by the Geddes Report, against the background of what bulk buying could do for you, we were behind. This being a family business there was not a lot of investment money. They did, religiously, every year, allocate a vote to the shipyard and engine works, and it was my job to prepare a list of proposed purchases. I do recall, with great annoyance, the time it took us to agree the purchase of a large Ram Borer for the engine works. It took me several years to obtain it. We had just taken out a licence to build Sulzer diesel engines [*Swiss-designed and manufactured, slow and medium speed marine diesel engines*], and for my money Sulzer had the highest quality products imaginable. It was more of a struggle to produce these with the inferior equipment we had. That is what finally won the day. In its day, the Ram Borer was a major event. They were not throwing money around.

I do not think that we had poor labour relations; this is before we joined up with Lithgows in 1970. One of the problems of the merger, and this must have been true across Britain, was that it meant that conditions of employment, wage rates and allowances were all different. I can not stress enough how much trouble that caused. Up to that point there had been a run of Labour Governments and the unions thought that they were on a good thing.

For the greater part of the 1970s Scott's and Lithgows continued exactly as they had before the merger, primarily because they built totally different products. The engineering for Scott's merchant ships and warships was my responsibility, and included building the main engines, and installing and commissioning all machinery and systems associated therewith. Scott's were [*conventional diesel electric*] submarine experts with familiarity with highly technical projects of high quality. Two examples were: a Liquid Petroleum Gas Carrier [*Gas Lion*] in1968, we were responsible for all aspects of the design-a very rewarding but taxing contract [*Scott's did not build another one*], and we built two dynamically positioned offshore oil drill ships [*Ben Ocean Lancer* and *Pacnorse I*] – largely engineering projects, which gave us an edge in the dynamic positioning field.

There was an awful lot of taking advantage due to the strength of unions at the time. In A.R. Belch's day we finished with one shop steward for every twenty men, and when you have 8000 employees that's 320 shop stewards. What caused me to climb up the wall was the fact that inevitably and most sadly, the guys who would take the job were those with a little more idea of their self importance than they really merited. The good guys, the good tradesmen, did not want anything to do with the shop stewards job. Inevitably, you were landed with a lot that was rubbish. On restrictive practices, we seemed to have bought that out three or four times. Every annual round of negotiations we would go for full flexibility, and we paid for it. The next year it was back on the agenda, and we paid for it another time. That went on and on. As an example, on demarcation disputes there was one that was pretty widespread nationally and that was between the electricians and sheet metal workers.

In a ship, if you were running electrical cables through it, you had to predetermine the route they would take. This involved perforated trays [*as a conduit for the cables*]. Who was to get the job of installing the trays caused endless trouble. The electricians argued that the trays were only there because the cables had to go in them, and that, therefore, it was their job. It was a typical example of hours and hours spent on negotiations. I used to find the sheet metal workers particularly difficult to negotiate with. They were just not prepared to move at all. In general, at Scott's we were not notorious for having a workforce that was out of the door at the drop of a hat. We had a fairly responsible body. When we got diluted with Lithgows that is when the fun started. There was even a struggle between the two sets of shop stewards as to who would be the top dog! That was never resolved satisfactorily. It alternated with the Lithgow stewards being on top and then our own.

One of the problems was because of the guaranteed four nights [*time and a quarter or half in hours worked in wages*] and a Sunday [*double time*] that labour got, and the fact that they got big wage packets meant that the attraction of becoming a foreman disappeared. That led to the fact that the people who got the foremen's jobs were not the best people. I did not envy the people who got them because they were up against a pretty rampant bunch of hooligans at times! That meant that your discipline was not all it should be, and it led to abuses of privilege and so on. The coppersmiths nominally made copper pipes, but to thoroughly confuse you, in warship building companies there was a sub-strata of demarcation, which said that the coppersmiths did all the piping on the engineering side whether it was copper, steel or whatever, and this caused a fight with the plumbers. These coppersmiths were the original Luddites to the nth degree. I privately used to burst out laughing when I thought about what we had to put up with from these people. There was simply no reasoning that would shift them. In the end they had to merge with the plumbers, and I do not think they profited from that.

A degree of flexibility did eventually develop among the workforce. Being a little bit parochial, one of the things Belch was good at was fighting one fire at a time. He was building huge ships [*250,000-deadweight-ton tankers*] at Lithgows, and what really got the Sheriff after him was their performance on the super tankers. They had shown great enterprise in getting a licence from Holland to build these ships in two halves, which were joined afloat [*in a specially constructed coffer dam*]. That meant huge amounts of welding. When A R Belch got into a jam, instead of saying that at least Scott's was running all right, he took fifty welders from us and sent them up to Kingston. This meant that in the end, both jobs were late. We were particularly anxious not to fall behind with three ships for Blue Funnel [*Ocean Transport and Trading*], a valued customer. It was in the middle of finishing one of these ships that a key section of the labour force was shipped off to Lithgows with the inevitable result. It was the fact that they were so desperate at Lithgows to complete one of those huge tankers that one manager commissioned a bulldozer, and bulldozed all the piled-up staging into the water. That's a bit of exaggeration in terms of what life was like normally. It was a measure, though, of what Lithgows got up to at the end of the day.

There was no separate marketing department. A.R. Belch was in charge of getting work, and spent a great deal of his time in London, usually three days per week. He had a permanent suite in the Charing Cross Hotel.

Did, British ship owners let the industry down? It depends on how you view the morality of the whole situation. For me, if you are running a business, you have got to make a profit, and then if your outgoings can be reduced then you make a bigger profit. There is no doubt that they took that route. They had no loyalty to anybody. We have to ask were they entitled to have any loyalty particularly when deliveries from a lot of shipyards were poor. Late delivery was a serious problem. It was part of the problem of working four nights and a Sunday. Nobody can do this every week and give four nights and Sunday worth of output. There was the Monday Club – the crowd that would not turn up for work on a Monday. There was the crowd that would go out for a drink on Friday lunchtime and could not perform on their return.

The Japanese did it differently on delivery. They said we will build a minimum number of these tankers, and they would spend a minimum of nine months with graduates planning every detail. Then they would be able to start production from a point they could go from, and were subsequently able to meet delivery times. I always found that we were on a hiding to nothing because you found that the owners would interfere to change details whilst the ship was under construction, and I believe British shipbuilders should have taken a tougher line here. The Japanese did not allow that. The whole flow of production went on unhindered with alterations being made after.

A criticism was that senior mangers were not all that well trained in the formal sense. I would agree with that. We did ourselves try to step up the technical side of things. In production, I do not think that the best choices were made on the management side. There was a particular case in point, the manager who had ordered the bulldozer, who was five notches up the management ladder from where he should have been. To be fair to A.R. Belch, and to Scott Lithgow in general, the last ten to fifteen years they had a very ambitious programme of under and post-graduate sponsorship and training, but sadly, they were not able to hold on to many of them. I found that in the Drawing Office we trained the Higher Nationals [*higher national certificate and diploma students, a step below degree level*] who were the Warrant Officers of British Industry, and we recognised that we could not find improving jobs for all of them. It did not stop us from doing it, and we made sure that these boys got the best out of training.

Personnel management was poor. Belch dominated, and he did not want anyone in their areas to take over his role. We had personnel managers who were basically "mouthpieces." In Scott's the personnel manger merely kept records of timekeeping, absenteeism and so on. Belch had a number of personnel managers who tended to be personalities rather than drawn from trained personnel. These people tended to be ignored, and grievances went to the next level above. The shop stewards just swept them aside. I think the only thing personnel managers did in the discussion between trade unions and management was to organise the meeting. That was all they were ever allowed to do. They were low calibre – I am not being personal – and the jobs they were doing were low calibre. They had no illusions either. They knew they were the pawns, the front guys who parried the first blows, and then brought it in for somebody else to deal with.

I think a plus side as far as A.R. Belch was concerned was that in the early days of offshore he spent a tremendous amount of time and effort in trying to get us into the market. This was long before we were really fit for it. The first thing we got was the two drill ships which were very successful. He was more interested in getting work for the Lithgow end with its huge steel handling capacity.

I think that the merger of Scott's and Lithgows was a shotgun merger; neither party wanted it; and I think that they were both correct. In my opinion it would have been better if we had stayed separate, and that is no disrespect to Lithgows. The fact that no problems occurred during the first six or seven years of the merger was due simply to the fact that we carried on the way we had been going before. We were building different products. We were building for the navy, and they were building for the large bulk market, and there was no cause for us to fall out. We were largely self-sustaining with the labour forces we had. When the trouble started, was when, as I said previously, they pinched labour from us to build those Blue Funnel ships. We finished up both in trouble. Michael Scott [*Chairman*] told me he had met Reay Geddes years later, and Geddes admitted that the Scott Lithgow merger was a mistake.

Was there an alternative to Government giving money to the industry? If you believe that the shortcomings of the industry were such that it could not match the Japanese in many areas, the effort could have been made to try to correct that situation. It was much more a palliative action that was taken. I think it had a tremendously political element in it, no matter which Government was in power. They were trying to see that unemployment did not get worse, and although I think everybody knew in their heart of hearts that once they got to grips with the Japanese they knew that they were really up against it. Charlie Connell made a suggestion that we should all get together and we should all agree on what the prices were going to be. That would solve the problem. Andrew Paxton [*Scott's*] asked who was going to invite the Japanese? They had still not got the message that the Japanese were the real enemy.

Was nationalisation necessary? I think that at the time it happened [*July 1977*] and the state the industry was in, if you were anything but fair and not a shipowner, shipbuilder or union leader, you would have said it was necessary, because it had the appearance of providing facilities with substantial support to try and correct these terrible variances between what we did and what the Japanese did. What changes did it [*nationalization*] bring? It provoked this strange structure at Headquarters, and there are terrible stories of chief executives lying poolside in Greece with their secretaries; they were people who had nothing to contribute to the industry. I do not think the personnel they brought in were satisfactory. Perhaps if we had people who were better able we could have done a better job. I do not think the industry would have survived better without it. We had to have something. The major advantages: well, the potential advantages were collective bulk buying giving greater bargaining power with suppliers, and collective technical strength. The disadvantages inevitably were that in certain people's heads the money was there for the asking. Because of the unions' collective strength under nationalisation they felt that they could get what they asked for. At the end of the day the unions got too much power. What happened to Scott Lithgow was as bad as what happened to anybody, and the union crowd have a lot to answer for.

Interview 3: Willie Ferguson, Technical Director, Scott Lithgow

I began as an apprentice in the Drawing Office of William Hamilton and Company in Port Glasgow in 1949. After gaining a Higher National Certificate I began my national service in 1955, and then returned to William Hamilton and remained there until they merged with Lithgows in 1960/1. I then moved to the estimating and design departments of Lithgows until 1967, when I became Chief Designer, and in the early 1970s after the merger, I became technical manager, and in 1976, Technical Director. In 1980, I became Technical Director to the Board of Scott Lithgow, but left the firm to join British Shipbuilders after Scott Lithgow became the first British yard to be privatised by the then Conservative Government in March 1984, when it was controversially sold to Trafalgar House plc.

In Lithgows Limited, which I was familiar with from 1960-1969, I would say that the strengths of the company fell into major categories such as good leadership and communications, technical strength, good facilities and investment in those facilities. We had established production processes with well defined functional organisation in the company. We had a good relationship with owners, brokers and ship managers, which helped to provide continuity of work for the same range of ships with the same owners. We had good training facilities for our apprentices. Being a family company, the family spirit was engendered throughout the company structure. We had weaknesses which, within senior and junior management were recognised. There was no real employee relations policy. We did not have a high-powered personnel department. The employee relations were a bit paternalistic, and whenever there were problems there tended to be a buy-out of the problem. Someone would assess the value and decide that the best solution was to buy the guys off. There was a poor graduate intake into production areas. Most of the graduates who joined the company tended to find themselves in the technical department. Perhaps, because of the family atmosphere, people were slow to accept changes, particularly in respect of management style and production processes. That may have been because of the age profile of the management, but, at least in the technical department, there tended to a younger age profile.

In the early to mid 1960s where the advent of computing processes was coming through we made some investment into that, but there was a lack of courage to change, and to accept new management styles and the production processes which were then emerging. I always felt that the big weakness was that there was no central planning function. There was a general planning function carried out which was on a broad base, and it was left to production management to set the detail and build strategy and planning for the ships. Ultimately, that became a big weakness in the company.

Scott Lithgow, which came into being in 1970, enjoyed strengths of having a wide product range of merchant and naval vessels that we could build, from dredgers, to super tankers and submarines. The weaknesses of Scott Lithgow were greater than the weaknesses which had existed on the old Lithgow side. There was no real integration of the companies. When Scott's and Lithgows came together they were two family concerns, and we had different cultures. Lithgows were the bread and butter builders, whereas Scott's were naval and liner builders. I think they gave themselves a higher profile than they deserved.

Ross Belch, who I knew very well and admired greatly, was the managing director of the joint company, but there remained at each end of the company an operational board, and the Board of Scott Lithgow consisted of representatives of each end of the company and some non-executive directors. They tended to dictate a policy which was then driven by the board members at each end. They interpreted that policy as they felt it was appropriate to their company. This meant that there was no real integration of the company and we solved our own problems.

Because of the different cultures in Scott Lithgow there were different production processes, and there was no real attempt to rationalise these. For example, Lithgows built with a single crane running up the centre of the berth, whilst Scott's built with cranes adjacent to the berth. Scott's heaviest lift was something like sixty tons, whereas in Lithgows Kingston yard ours was eighty-five tons. Because of the two Board structures in each end of the group, there was poor functional organisation. There were attempts made to have common purchasing to get the benefits of bulk buying, but things like planning, quality control, production control, personnel, wages and general accounting remained independent at each end of the company.

We still had good communications through Ross Belch. We did have company seminars, and were encouraged to speak our minds. In the latter days of the 1970s before nationalisation, there was an attempt in the company to change and bring more strength into production from the graduate side, but it was a bit too little too late.

One of the weaknesses of Scott Lithgow was that there was massive resistance to change. Scott's were a monolithic empire in their own minds and to get them to change the habits of over 250 years of shipbuilding was almost impossible. By the time Scott Lithgow came together in 1970 there was a great tightness of money all round because of the state of the order book, and the state of cash in the company. So there was a lack of investment for new products and processes. We were not really setting down the way we had before looking ahead and seeing what investment was required. I do not think that there was a proper investment programme from 1970 onwards.

In the late 1960s when the Geddes Report came out, I did not really think that "big is beautiful" was a great idea. The idea of integrating with Scott's did not really appeal to me as a first solution.

People like IHI and Mitsubishi in Japan were going off and building new shipyards in Greenfield sites. We [*Scott Lithgow, in conjunction with A&P Appledore, Consultants*] helped Hyundai to build their shipyard in South Korea. We trained their men. They spent money on it as if it was going out of fashion. Whereas our modest conversion of the Glen yard to build a super tanker, was very minor in comparison. When you think of the achievement of building super tankers in two halves at Scott Lithgow; it was necessary because we did not have the space in the shipyards or capital to expand into a facility that could build them in one part. Whilst it added cost, it was not by an enormous amount, but it kept us at a disadvantage.

By the mid 1970s the industry was bankrupt, because every shipyard lost money. The way in which Scott Lithgow was surviving was because it had a balance between merchant and naval work which was profitable. Then in 1977, nationalisation came along. Most of us at Scott Lithgow realised that as a company we were hanging on by our fingernails, and the industry was doing the same, so that in that background nationalisation could have been a saviour.

Having looked at what happened in the 1950s, when coal and steel were nationalised, I felt that with all the experience that government and the Department of Trade and Industry had at nationalising industries, the shipbuilding industry nationalisation would form an almost perfect model. That was a naïve vision. When nationalisation took place in July 1977, I was amazed to find that there was not a very good plan at government level, and most of the work that Graham Day had done on the organising committee before leaving, had not been accepted by the government. They were still debating by themselves. The result was a paper nationalisation, and there was an organisation called British Shipbuilders Headquarters set up which had no teeth or muscle, and had no shape or form. It was not until 1984 when I joined them, and spoke to people who were involved in that process, that I discovered that they were really picking their way through a treacle field. The guy who was the company secretary told me that on day one, they did not even possess a list of the companies that were to be nationalised! He was writing to companies like Scott Lithgow asking them to tell him what companies were in the group and what the assets were. Nobody knew, at that point, what the corporation consisted of! Why were all the recommendations of earlier consultants reports not recognised in the nationalised corporation?

In Scott Lithgow, weaknesses in labour relations became the cornerstone of many of its problems in the latter period from 1981-1983 [*by which stage, Scott Lithgow had basically became an offshore specialist building two semi submersible drilling rigs, both of which were years late and both were subject to huge financial penalties*] before it was privatised. People in the shipyard had grown up against the background that a bit of militancy could overcome negotiation or it could lead to something else in their hands, not necessarily an increase in wages, but allowances. For example, people in skilled groups would protest about something, and to buy out the problem an allowance would be given to them. People began to accept allowances as being a part of life. I think the policy of buying out problems became the cornerstone of the big let down form 1981-1983. That period was dreadful.

In 1980, British Shipbuilders adopted a divisional structure. The new British Shipbuilders chairman Robert Atkinson placed Scott Lithgow in the Offshore Division. Offshore standards were considerably higher than mercantile standards and all the welders had to be re-coded. Immediately that brought demands for higher pay. By 1982, just before the end of Atkinson's reign, the cutbacks started. The Corporation was indeed suffering losses [*Scott Lithgow was its greatest loss-maker*] and Atkinson was told by the Conservative Government that he had to cut back and consider privatisation. I had to admire him for refusing to privatise the warship building yards; albeit, they were getting their profits from the public purse, but that was helping to support the less profitable parts of the industry in a bad market. Atkinson left in 1983, and was replaced by Graham Day. In 1984, his first Corporate Plan was very interesting; he created an empire within British Shipbuilders which had composite shipyards building fleet support vessels, warships and merchant ships. He kept the merchant ship division and the engine building division. He began to dismantle the Offshore Division which had not been very successful [*it was a complete disaster*] and he was told by government that he had to privatise the warship builders. So his plan envisaged a situation where the offshore and warship divisions would have gone, but he would have had the composite builders, merchant yards and engine builders, and that they would remain nationalised. He believed that this was the lowest critical mass that you get to and remain competitively viable. He set up a central structure of management who were accountable.

When one looked at Day's Corporate Plan in the middle of 1984, you felt quite encouraged, but by the autumn, Day had been told by government that those shipyards which were classed as composite, and which could also build naval vessels were also to be privatised. Principally these were Swan Hunter on the Tyne, and Cammell Laird on the Mersey. That knocked a massive hole in Graham Day's plan since the composite division disappeared. Thus the critical mass of the operation was reduced, and we were left in a situation where merchant and engine builders were left to support the monolithic structure at Headquarters, which had eventually to be cut down.

By 1987, British Shipbuilders had a new Chairman, John Lister who wanted to consolidate a minimum critical mass of the yards which were still left in the corporation. He produced a plan in 1987, to that effect. But in September, the minister in charge, Paul Channon, who was sympathetic to the plan, changed office and in came Lord Young who was supported by Kenneth Clarke in the House of Commons. Lister, at first, found it very difficult to get access to these men because they were re-organising the Department of Trade and Industry. Finally, when he got to see Clarke he did not want to support Lister's plan, because, in his view, the support costs were too high. He gave Lister instructions to pursue the avenues of privatising the remainder of the corporation. John Lister went public in saying that this was not what he had been employed for, and argued it out with them. Nevertheless, at the beginning of 1988 he was given the final instruction to privatise, and told to get on with it. This led to the total privatisation of British Shipbuilders. Govan went in 1988, sold to the Norwegian based Kvaerner Group, followed by Appledore, Ferguson and others. Sunderland was different; there was a lot of hot air on the subject of North Eastern Shipbuilders Limited [*an amalgamation of the River Wear yards*]. Eventually NESL was closed and the area designated an Enterprise Zone in order to attract European Community funds.

Nationalisation created a tremendous opportunity, but it was badly managed.

Postscript:

In March 1984, Scott Lithgow, which at its height had employed 9000 employees, mostly from the Inverclyde area, was privatized under the ownership of Trafalgar House Plc at a bargain-basement price, with British Shipbuilders picking up the bill for redundancies. Under Trafalgar management, the workforce rapidly contracted, and local management was unable to bring any substantive work into the yard. Trafalgar belatedly tried to re-enter the submarine market without success, and continued work on two large, semi-submersible, dynamically positioned offshore drilling platforms (already years late under Scott Lithgow management) for British Petroleum and Ben Odeco, both of which were eventually delivered even later than earlier envisaged and absurdly over the contract price. By 1990, what was left of Scott Lithgow was placed on a care-and-maintenance basis, the land at the Scott end of the business was acquired by the Scottish Development Agency and the remaining land at Port Glasgow was eventually sold in 1996 to Clydeport. Only one small shipyard, Ferguson Shipbuilders, once an integral part of Scott Lithgow, remains in the cradle of steam navigation on the Lower Clyde.

Interviews
Upper Clyde

Interview No. 4: Sir Eric Yarrow, Chairman, Yarrow Shipbuilders, Scotstoun

I served my apprenticeship with Weir's of Cathcart, and I was at Glasgow University, but when the war started I had not completed my full training. I served six years in the Army mostly in the Burma area, and was demobilised with the rank of Major. I then spent some time in a number of shipyards and engineering works for the best part of a year before joining Yarrow's. I started in a relatively junior capacity [his father and grandfather were both chairmen of the company] and made my way up through the company becoming Managing Director, and later Chairman and Managing Director.

I think really one of the three reasons for the success of Yarrow Shipbuilders was the fact that we concentrated on one particular type of ship, which was the naval ship or the highly sophisticated ship. One of the reasons why the shipbuilding industry declined was that some companies tended to do a too great a variety of ships in one yard, and I do not think that is very effective. It is very difficult to build a frigate or a destroyer alongside a bulk carrier or a tanker with overall financial success. If someone asked us to tender for a cargo ship or a tanker, we would not look at it. I think a second reason why we proved successful, was that we did plough a lot of money back into the business to keep it up to date. We were the first yard to have a covered building berth of any size, and we spent considerable sums of money every year on the company. The third reason is that we did take a lot of time and trouble over labour relations. We thought it was very, very important. At one stage in the history of Yarrow's the shop stewards actually convened a conference to pay tribute to the relations between the management and unions.

The fact remains that the profits Yarrow's attained prior to nationalisation were to a significant extent due to building ships for overseas customers, not from the Ministry of Defence. At one time, there were fourteen warship building yards in the UK, which was far too many. During the 1970s Lord Carrington made it clear that the Ministry of Defence should concentrate warship building [*submarines and frigates*] in just three specialist yards, of which Yarrow was one.

Much against our wishes, we had earlier been ordered by the Government to join Upper Clyde Shipbuilders [UCS] in the wake of the Geddes Report. Its no secret that I had a letter from the Controller of the Navy at the time making it clear that political pressure might be put on him to deal only with those yards which had joined groups. Having heard that, we had no alternative but to join the group, but we were a public company owned by shareholders-this is an opportunity to say that sometimes people thought that I owned Yarrow Shipbuilders, but I only owned one percent – and I had shareholders to consider. We went into UCS to the extent of fifty-one percent owned by the Government and forty-nine percent by Yarrow. It was not a satisfactory arrangement, and it did not prove successful. In fact, we were damaged. I have never been a believer in the bigger you are the better you are, and I had doubts about UCS from the start. What the government tried to do was put five companies together, three of which were making losses, one of which was making ends meet, and one, Yarrow, which was profitable. You can not do that and hope to make an overall profit, no matter how much government funding is agreed [*£20 million to UCS before 1970*]. I was also, and still am, a believer of supporting the strong and not propping up the weak. The strong should be made better and more competitive in world markets.

What we really desired was genuine competition. In building ships for overseas countries we can not be competitive if some countries are giving subsidies. We were in competition with the French, the Italians, the Finns, the Dutch, the Swedes and the Spaniards in Europe alone. A great mistake was made by the government in thinking that if the industry was nationalised, the employment is more assured, and there would be less redundancies and closures. But that just is not right. I saw Geddes just after his report came out and I told him then that I did not think that his proposals were going to work. Then, of course, later on there was the sad story of nationalisation. For some yards, it was a lifeline. If nationalisation had not come about they would have been sunk without trace. We [*Yarrow*] fought it up to the hilt. It was very unfortunate that on one occasion the Bill only got through the House of Commons by one vote. To make matters worse, the amount of compensation Yarrow Shipbuilders got was grossly unfair. I said at the time, and I say it now that it was "robbery." The compensation terms [*at 1974 values, when the Labour government first announced its intention to nationalize the shipbuilding industry*] were absolutely disgraceful, and the government knew it. When the Conservatives came to power [*in 1979*] they were not prepared to have retrospective legislation, although they had also fought the Bill up to the hilt when in opposition. I can only give very rough figures if my memory is right. After a haggle with the Government, and before going to Strasbourg to the European Court of Human Rights [*he and Sir William Lithgow lost their respective cases for additional compensation*], Yarrow were offered £6m. Yarrow Shipbuilders when nationalised poured £34m into the British Shipbuilders kitty, and GEC bought the company for £36m. The government thus got some £70m while paying out £6m. Even taking into inflation into account, such a settlement was totally unjust.

As far as naval shipbuilding [*for foreign account*] was concerned, there developed more and more overseas competition. One reason was that we provided know-how for certain countries, which were incapable of building warships without our help. If we did not do it then somebody else would. To that extent, when countries began to produce their own warships, the market was reduced. On the other hand the Ministry of Defence wanted to go to shipyards which they were absolutely certain would give top priority to their ships.

On marketing, we had a good sales team, and in addition I tried to visit as many countries as I could. We kept a chart of countries which had been visited over the years, and any particular country which had a requirement for naval ships we would try to visit every three years. We preferred to do our business by going out and about, rather than sitting at our desks and telephoning and writing letters. The important thing is to see people on the spot. Right after the war I made a trip around South America, and I made frequent visits to Thailand and to Malaysia with whom we did business. There was Marketing Manager who had a small team, but I think there was a lot to be said for the Chairman or the Managing Director going overseas. It may not seem right, but it seemed to happen that if the Chairman or Managing Director was there he could possibly open doors that would otherwise be closed.

In terms of labour relations in the industry, I think a lot of blame was put at the feet of the trade unions for having too many trades in the one yard. I would not, by any means, attach all of the blame for what happened in the industry to the trade unions. The fact remains that they did change slowly, and we were up against countries that had maybe one or two trade unions to deal with. In UK shipyards, there was a multiplicity of trades unions. I think that was definitely detrimental to our competitiveness, but that was not the only reason. It is just a pity, in retrospect, that the trade unions did not change more quickly.

Interview No. 5: Dr. Brian Baxter, Shipbuilding Director, Yarrow Shipbuilders

I started as an apprentice in the Royal Dockyard at Sheerness, Kent, and was sent to the Dockyard at Hong Kong. I joined up there, but when the Japanese captured it, I became a prisoner of war for four years. At the end of the war I returned to Britain and became a student of naval architecture at Newcastle University. On graduating, I joined the Royal Naval Scientific Service and served at Rosyth. I subsequently left Rosyth to take up a lecturing position at Newcastle and taught there from 1950-1957. I then became the chief representative of the French classification society, Bureau Veritas for five years. In 1962 I joined Yarrow as Shipbuilding Director, and stayed there until 1982. From 1980 to the present, I have been the Visiting Professor in Naval Architecture at the University of Strathclyde.

When I started at Yarrow in 1962 everybody was our competitor, because there were ten yards at least in Britain who were building warships: John Brown, Stephen, Denny, Scott's, Fairfield, Swan Hunter, Cammell Laird, Harland and Wolff [*all mixed naval and mercantile yards*] and others. All were building Leander class frigates, so it could be said that we were in competition with all of them. Yarrow believed and said, "none of you are as really dedicated as we are." The possible exceptions to this were Vosper Thornycroft [*on the south coast of England*] and Denny [*at Dumbarton*] who went out of business in 1963. They had the same sort of feeling about the Royal Navy that we did, and in many ways they were similar firms. They were all strong family firms formed by strong characters, namely, Sir John Thornycroft [*on the Thames*], Sir William Denny, and Sir Alfred Yarrow [*on the Thames*], and they all restricted themselves to the same sort of small warships. We were not like John Brown, Stephen or Fairfield. We were going to build small warships, and that is what we told the Admiralty [*the department of state responsible for warship construction and repair in the UK up to 1964*]. We not only did that, but we also made a point of asking, what's the size of the largest frigate you are going to order over the next decade? We deliberately built our covered berth and our fitting-out basin to accommodate that size.

Naval building and naval requirements always came first, as distinct from some firms on the Clyde who built warships and merchant ships, and tended to play one against the other. At Yarrow's whenever we considered plans to alter, change or better our facilities, we would unofficially submit the proposals to the Controller of the Navy first of all. Yarrow benefited from the fact that because they were engaged one hundred percent in Admiralty work, some of the overheads were borne by the Admiralty. If we were building facilities for a particular class of ship then clearly we thought we should be favoured, not from a financial point of view, but from the point of being able to tender for those ships. I then believed that our expertise and experience would enable us to obtain the orders because we would be selling at a lower price.

The major strength of Yarrow was that they had decided and kept to the decision to concentrate on the design and building of moderate size warships. I stress the question of design. For a long time we believed that the number of graduates in Yarrow's, particularly on the naval architecture side, was much higher than in other yards. Let me broaden this point to say that in general terms throughout the industry the naval side had a higher calibre of design office personnel than the merchant yards. Vickers, Vosper's and Yarrow's at one time must have had sixty percent of all graduates of naval architecture in the country. In comparison, the merchant yards had few. They did not concentrate on the higher qualifications which we thought were necessary. I think that because they did not have highly qualified people they tended to perpetuate a system of the shipbuilding industry of what I shall call the old fashioned syndrome. The belief was that the only way to become a competent shipbuilder was to spend thirty years going up the ladder. They thought it was not necessary to have or to acquire wider qualifications in shipbuilding or engineering. This tended to restrict the ability to develop new and better ideas. This stood in contrast to the Dutch, the Japanese, the Germans or the French who were firm believers in the notion of high qualifications. This situation generally worked to the detriment of the UK.

I think that the Geddes Report was an extraordinary good report. I also believe that his suggestions [*actually his committee's suggestions*] for overcoming the ills of the industry, had they been followed, would have cured many of its problems. The fact is that his ideas were good, but were not followed by the mass of shipbuilders. I think it was probably because of the idiosyncrasies of individual yards. The shipbuilding yards of the early 1960s were dominated by strong-minded boards who never believed that the solution to the problem was either amalgamation or co-ordination in a big way. They believed that they were capable of getting their own companies out of the mire. They were not too interested in the plight of other yards. I think that many of them did realise that they were in deep trouble. However, many shipbuilders, even at that time, were optimistic in the sense that many of them had been brought up on the idea that the shipbuilding industry has troughs and peaks [*in demand*]. No trough would last that long that they could not get out of. I think many felt that the trough was really deep, but they would rise up again to a peak on the other side.

When I started at Yarrow's I said to the chairman that I had never really been a shipbuilder before and that I would like to visit Swedish, Norwegian, Dutch and French yards, so that I could bring my ideas up to scratch. He said it was no problem, so I visited the well-known yards in those countries. A little later on I said it would benefit the company if I visited Japan, and had a look at their shipyards. I suggested that it might be possible to apply their production and design methods to our own yard. This is what I did. I think that the criticism that the British yards were fairly outdated in the 1960s is fairly well justified, particularly as far as labour relations were concerned.

I can remember doing a survey of the amount of equipment in money terms that was available to each man in the British shipbuilding industry, and we were far, far behind. That is not always a strict way of measuring efficiency; I can remember one of the worst yards in the country for physical equipment was one of the few places that made a profit, Burntisland Shipbuilding in Fife, because they had the ability to make the men work hard. Because of the naval dimension at Yarrow there was never the anxiety of the workforce and antipathy between management and men which may have arisen in other yards that were on the point of closure. Yarrow's was somewhat cushioned by that, but having said that, I do not really believe that the level of compliance with good working practices, and the solution of demarcation problems was higher in Yarrow's than it was elsewhere.

I would not say that Yarrow's was in advance of any of the other yards in the interchangeability of workmen, or in the solution of demarcation problems for which we paid fairly handsomely to have solved. As soon as one particular problem was solved, another one arose. The comforting thing in Yarrow's was the better relationships which did develop came about through the understanding, rightly or wrongly, that Yarrow's were never going to close down.

The problems of actually organising and agreeing workloads and work practices were formidable in British yards, compared to elsewhere. I think that, to some extent, the quality of production management generally was poor, and here I return to the point about qualifications. All of the junior supervisors were promoted workmen. Many of the assistant managers, and a lot of the managers were recruited from a craft basis. They knew their job, they did it well, but it was extremely difficult to separate their mental attitude from that of the normal hourly-paid. The number of production personnel of all yards in Britain who had any senior qualifications, was minimal.

I believe that although the nationalised industry has been much maligned, British Shipbuilders worked very hard on their training and education programme. They set up a training and education company of which I was a director. They made certain that scholarships and money were made available for bright young employees. I believe that this was one of the very positive aspects of the nationalised industry. On the craft side, Stephen's was a very good yard from the training point of view. In Yarrow's there was always a hard core of very good men who you could depend on to turn up and do a first class job. This central core would never move.

Generally in the industry there was not the equipment available to lead to higher levels of productivity. There is also the question of the British character. The British are lazy compared to the Germans or Japanese. This was very worrying for the whole of the British industry, and certainly in the 1960s, all sorts of organisations were set up to try and find out why it was that the levels of shipbuilding productivity were so low. You could find many of the explanations, but there was always a final factor of ten to fifteen percent poorer productivity for British companies, for which there was no explanation.

I can remember in 1979, when John Parker, who at that time was Deputy Chairman of British Shipbuilders, decided he would try and measure the productivity as far as working hours were concerned in the British yards. On a certain day, every man's output would be measured. The results showed that out of an eight-hour working day, three hours and five minutes were not productive, for one reason or another. It does not matter whether the workers were bone idle or lazy, or whether there was no crane available, or no welder or other support. The fact is the average for the British shipbuilding industry was that out of an eight-hour day, three hours and five minutes were wasted. I know that time was wasted every morning in starting work, in the famous tea break, in getting from the ship to the place where you had tea, and getting back again. Time was wasted stopping early at lunchtime, starting late in the afternoon, and stopping early again at the end of the working day. When I went to a German yard I found that they had solved the problem of tea. They had installed in their yard a number of tea machines, and every man had the ability to get a cup whenever he wished. When I started in Yarrow's I said to the Shipyard Manager, get tea machines and tell everybody they can have tea when they liked. And the Shipyard Manager said, very politely, "that this was a daft idea, it will never work." I did it nevertheless, and when we had the machines installed it was a dead loss. It worked for about a week. Thereafter, what happened was that when nine o'clock came there was a queue as long as your arm. By the time 9.20 a.m. came they had not got their tea, so that was that!

In marketing terms, British yards were generally very poor. In the 1950s and 1960s very few companies had a powerful sales manager, and very few, if any, had a sales director. What happened in most of the yards was that the technical people did most of the selling. In many companies, including Yarrow's, it was the chairman, managing director and technical director who were considered to be the best qualified to do the selling of the company's products. It was only at the end of the 1970s when marketing and selling was regarded as being worthy of a senior appointment. We got the Naval Director of Sales, and we recruited him as Sales Director in the 1960s with an office in London. He kept details of our designs. Our chairman was always very good at maintaining relationships with overseas naval staffs.

What we tried to do in the early days was to forecast technically what was wanted. For example, you looked in *Jane's Fighting Ships* and picked ten countries, Malaysia, Pakistan, Thailand etc., and found out what they had in 1961 or 1962, how old they were, and then come to the conclusion that in three, five or ten years time they would have no frigates or submarines. We would get in touch with them, and say that we had a brand new design for a frigate, how would you like it, and can we come and talk to you about it? That was the way we tended to do it, rather than send out somebody to India for example. By and large, we got orders from these countries. In this way these people in one sense did not know what they wanted, or what the advantages of a particular design were; they were not stupid, but our approach had advantages over going out there and saying what particular frigate would you like?

After nationalisation, British Shipbuilders developed and maintained a very strong marketing system, and in fact opened an office in Hong Kong, Greece and in the United States, then produced beautiful standard designs. Unfortunately, for British Shipbuilders their first decade between 1977 and 1987 coincided with the greatest depression in the world shipbuilding industry this century. Various governments gave £1.8 billion to the industry over a period of twelve years between 1977 and 1987. That's a lot of money, and so it is hard to say that enough money was not given. The real question is what happened to that money, and that takes us back to individual firms and managements. I calculated by finding the difference between the rates of pay of a general industrial worker compared with shipbuilding workers, that over a period of eight or nine years, much of the money that was given to the industry was reflected in increases in shipbuilding rates of pay, almost mathematically.

I am in no doubt that nationalisation was necessary for the industry's survival. I never like to moan and whinge about unfair competition. Shipbuilding in Japan in the 1960s and 1970s, and today, is and was very much a high profile industry. If you were a shipbuilder then you were a very proud man. The average level of their workman in terms of education was something like two or three years above us. Leaving aside their motivation as Japanese they were on a higher level, the number of graduates they employ is very much higher than here. Everybody in the Drawing Office was a degree man. And then you have the Japanese spirit that Japan is the supreme industrial country. They work under a highly disciplined organisation as far as their work is concerned. This is true also of the quality of their work. They are backed in various ways by the Japanese Government, not in a direct way, but in an indirect way. The fact that we have not developed in our shipbuilding industry in this pattern makes it very, very difficult for the British industry to achieve the heights that Japan has attained.

Interview No. 6: Bob Easton, Managing Director, Yarrow Shipbuilders

I served an apprenticeship as a marine engineer with Fairfield, and in 1951 I joined Yarrow in a junior managerial position. In 1965, I was appointed a director of Yarrow Shipbuilders which was set up at the time of UCS. When Yarrow was extricated from UCS I was deputy to the managing director, Ernie Norton, before later becoming managing director in my own right.

The major strength of Yarrow's was that it concentrated on specialised fields of construction. When I arrived at Yarrow's one of our strengths was the construction of small river craft for estuarine waters and inland lakes. Specialisation had permeated all through the scene, and we were known specifically as destroyer builders-from there we eventually moved to frigates. Yarrow's to me were always a group of highly motivated professionals. It was always a question of looking at other things we could tackle, never being afraid to try new avenues, or walk into the unknown. There was that certain spirit of get up and go. That; coupled with the fact that we had a merchant and a warship side to our business. Years before, in the days of Sir Harold Yarrow [*Sir Eric Yarrow's father*], we had entered the land boiler business; that was where you had the not all the eggs in one basket syndrome-the shipbuilding could be down, and power stations be up, and *vice versa*. In the late 1960s and 1970s however, there developed a surfeit of boiler plants.

We lost something when we reluctantly left the land boiler market. Then again, in the 1960s we had found ourselves with an overseas naval market where we were very successful in winning new orders, and at one time we were building ships for eleven different nations-two thirds of our output being for the overseas market. The customers appreciated that when Yarrow's quoted a price and a delivery date you got the ship at the price agreed and on the delivery date agreed-and the quality was excellent.

In the period between 1969 and 1972, Sir Eric Yarrow, Norton and I took the company from near bankruptcy to a position of strength and back to profitability. In the course of that we borrowed £4.5 million from the Ministry of Defence to keep us in business. We brought back our man hours to normality, and paid back the loan with interest, and as such owed no favours to anyone. From that point onwards, we proceeded to enhance the labour force from 1600 people in the 1960s, to a figure of 5600 people by 1977.

On nationalisation, we were given all of the advice about what would happen with British shipbuilding, namely that there would be a central core, and that all units would be autonomous, and live or die as separate profit centres. Headquarters was stated to comprise no more that ninety-nine people, but by the end of the day Headquarters was made up of a vast amount of people, and the idea of autonomy had disappeared from quite a number of companies. Yarrow's always maintained much of their autonomy.

Another strength was that there was an aggressive policy to market at the time and we found ourselves in the late 1970s involved in Type-21 frigates for the Royal Navy. From Type-21 frigates, we went on to build Type-22 frigates, and were proud to lead the design and build of all three batches of the class. They were like light cruisers in size. Our capacity was so taken up during that period that we could not indulge so freely in the overseas market as we would have done in the past. It so happened, however, that the overseas market was cyclical, and when we were on the crest of Type-22 building, the overseas market was in the doldrums. Nevertheless, in that same period, we also won six Logistics Support Ships for the Imperial Iranian Navy, which filled our capacity.

The major weaknesses; particularly of the British merchant shipbuilding industry, was a failure to recognise a role in the overseas market. There were a number of up and coming countries like the Japanese who were followed by South Korea and Taiwan. The method of building ships had to change, and technology had to be improved. Even today, some yards still build ships in what look like construction sites. There was also a failure to recognise that the trades in the world were changing, and that people were not being trained for the task. You had hidebound working practices which still existed from the 1930s. Concessions had been won, which people were frightened to lose. There had been a failure to recognise that one of the concessions-restrictive practices-was, frankly, totally outmoded. Nations abroad had learned the lessons; you could say the answer was that it lay partly with the workforce, and partly with the management. In this yard today, we have realised that it is no longer a "them and us" syndrome. There is a definite feeling here that we are in this together, and we have to find work between us if we are not to join the illustrious list of famous shipbuilding concerns that did not learn the lesson. The younger folk of today have awakened to the fact that a number of things we have done in the past are outdated. In terms of flexibility, working practices and composite squads, we have picked the brains of oil companies, and the Japanese, and added Clydeside common sense.

British Shipbuilders as an entity had a chance to succeed in its initial stages. The concept was reasonable, the attitudes were right, and a number of people initially had the right idea. If it had gone along that way, and utilised the availability of existing professional skill in the industry, it might have been more successful. The world scene was changing, and world politics as such have a bearing on other countries as well as on the UK. Unfortunately, British Shipbuilders became a bureaucracy. The autonomy of units did not work out the way it started out. The individual units should have been left alone to work on the premise of competitiveness. British Shipbuilders lost its way in a similar fashion to the situation during Upper Clyde Shipbuilders in the late 1960s and early 1970s. UCS was an attempt to rationalise totally disparate units into one cohesive whole. The constituent parts were so disparate that it was quite unbelievable that anybody should want to try. You found yourself with people being brought into the industry who lacked knowledge of the industry. The necessary leadership quality and shipbuilding experience was not there. The attitude became simply one of pumping money into an organisation where the speed of the convoy was the speed of the slowest ship. That was no way to compete against international competition.

In the period 1977 to 1985, [*during the nationalization period*] we always had a programme of modernisation and refurbishment of the shipyard. The final outcome of that was when we built our new frigate module building hall. That hall is only a production tool, and is in no way a full answer to the problem. The problem is people and systems, and the building itself is part and parcel of the overall chain. We have spent a considerable amount of money – about £55m – over this past twelve years on capital development, and we now have our latest project, the frigate building hall, where all the latest ideas of construction techniques and computers are put into effect. We now stand foresquare with the world's best warship builders as far as construction is concerned.

The British are among the finest shipbuilders in the world when the chips are down. Regrettably, there has been a history of too many lost opportunities.

Interview No. 7: Sandy Stephen, Managing Director, Alexander Stephen Shipbuilders and Engineers, Linthouse

Stephen's was a family business. I was the seventh generation of Stephen's. In spite of my father urging otherwise, I became a shipbuilder, and having been in the navy and at university, I went to Alexander Stephen's for a spell. Then I worked at Burntisland and at William Hamilton in Port Glasgow before coming back to Stephen's in a junior managerial position. I went on to deal with the sales and marketing side, and was very briefly managing director in 1967-1968 before the merger with UCS. After that, my position with Stephen's was as a non executive, and although we were not in shipbuilding, we did run our engine works and ship repair sides for another eight years.

As far as the strengths of the industry in general, I am not sure that there were very many, except that there was a great rapport between the shipbuilders and the ship owners. With hindsight, we obviously got ourselves into a hole and we could not dig ourselves out of it. The weaknesses went right back to Victorian times when a labour structure was set up, which did not really allow for any form of modernisation or alteration. When technical developments arose, the system was such that the industry could not modernise.

Demarcation was always there. It was always a problem fitting in between one tradesman and another. A demarcation problem would come to a head for perhaps some disconnected reason. If there was some unrest somewhere, all any department needed to do was fix their eyes on demarcation, and say that this chap next to us is doing our job, we want more of it. It was always available as a battleground. It was only occasionally that it was taken up. Normally, there was a lot of sensible demarcation. Trade union officials were politicians. Though elected by their members they had to deliver the goods. They had to make sure that their members did as well as other people. Their job was not to start with the interests of the industry.

I think our most devastating experience was in the early 1950s when we were probably leading Britain and most of the world in pre-fabrication. We realised that it was stupid to split the work between shipwrights and boilermakers. My uncle John pioneered the combined team. We suggested that it should have sixty percent boilermakers and forty percent shipwrights, or perhaps, it was the other way round. Each of the trades wanted to get a higher percentage in the numbers working. This resulted in a six month strike, which virtually brought our company to its knees. We were quite close to bankruptcy. There was no way either union would give way. In the end we got the compromise which was almost exactly what we had proposed. What we really wanted to do was to do away with all the trades and have steelworkers who would work steel in the way that joiners would work wood, and do any job connected with steel. You could have fitting-out people who could do any job connected with the fitting-out of ships. The answer might have been to have one class of shipbuilders where people were trained to do the whole range of tasks. It was the system that we were locked into that was the problem.

As a company, Stephen's had more graduates per 1000 workers than any other shipyard. We took on a lot of graduates compared with some British shipyards who had few or none. The people with degrees were not always better than those without. I was rather surprised on one occasion when our Naval Architect said to me that after the age of thirty it did not matter a damn whether a man had a degree or not. You are relying on ability. In those days a lot of good people did not have degrees. I think our senior management was as well trained as the average of British industry. One of the problems in the British shipbuilding industry was that the potential to manage was not that great, because you could not make progress. You were stuck. I think that a lot of young men who worked in the industry left and did something else. They could also see there was no chance of it being a growth industry. We were a training ground for young graduates; we had no real shortage of able people.

We had a very good customer base with companies that we built for year after year. It was also a weakness that when the [*passenger*] liner industry in Britain got itself into trouble, our market disappeared to a large extent. Apart from cargo liners we also built warships, which in a way caused us certain problems, but were very worthwhile. We specialised in high-class cargo liners. Before the war we had built passenger ships, but when the aeroplane took over, that market all but disappeared. We built our last passenger ship in the early 1950s. It was a market change rather than us losing our share of the market. I think that we also realised that the cruise ship was going to come along, so we kept our passenger shipbuilding facilities trying to capture that market. However, it did not really come along until the 1970s and 1980s. We were really ahead of the game in that sense. We specialised in high-class cargo liners, insulated ships, in particular refrigerated meat, bananas and so on. As far as the British competition was concerned our rivals were Scott's, Hawthorn Leslie [*Tyne*], Cammell Laird [*Birkenhead*], Fairfield, and, to a degree, John Brown. We tended to be in competition with them for almost all the ships we built.

Generally there was a lack of investment; but one the reasons for this, was over-manning. There was undoubtedly less incentive to invest money in machines knowing that they would be over-manned. Rather than buying new machines, the prerequisite was to ensure that the ones you had were manned properly. There was more to be gained that way. I think that the real problem was high costs. In the late 1960s we were finding that the Japanese were quoting ships at two-thirds the cost we were. However good your quality, however good your relations with customers were, it was very, very difficult to compete under those circumstances. Late delivery was not really a problem with Stephen's once we had caught up from the 1950s. Ships were late then, and the reason was that you could never get the material on time.

I think we did marketing inadequately. As far as our company was concerned we set up a marketing department after we had taken over Simons and Lobnitz [*an amalgamation of two world famous Renfrew dredger firms*]. We did search around the world very much harder to get orders, not that it produced a tremendous amount of benefit. It was probably too late by that time.

You could not blame British ship owners; they had to run their own businesses. They, in the 1950s were certainly a lot less inefficient than we were in running their business. They could make money [*during the Long Boom in international trade*] standing on their heads. They got into ruinously expensive habits, and they experienced just as big a decline in their business as did shipbuilding. The cargo liner market took the biggest hammering of all. I think that the owners we came across were extremely loyal to the British shipbuilding industry. They came to us on occasions when, if I had been a ship owner, I would have gone to Japan.

Japanese competition came on from the mid 1950s. At that time we were of the opinion that the quality of Japanese ships was less good. It was only in the 1960s that we found that there was nothing to choose in terms of quality. After that, we realised that we would have to compete on equal terms. In the early 1960s I saw no way to compete against them as an industry. It was a very depressing few years that I had urging people to do better, when I knew that the task was really beyond us. We [*Stephen*] did not lose many orders directly to Japanese competition. I think that they tended to go for the giant tanker and bulk carrier. We were in reality competing with the cheaper end of the British market, and decided not to specialise in large bulk vessels.

We always regarded foreign competition as unfair. This was the only way we could get our own government to help us by saying that competition was unfair. It was, of course, very difficult to prove. I think we did feel that the Japanese were being subsidised, but not necessarily in the same way we were. In our heart of hearts, I think we knew that the Japanese were building ships cheaper than us. We would then have needed a substantial government subsidy if we were going to keep our yards open. We found that there was little difference between Labour and Conservative governments. It was not a party matter. We needed a massive subsidy, and quite frankly I never saw the government coming up with that sort of money. The government saw its function as trying to regulate the decrease in size of the industry. They realised that the industry had to shed enormous numbers of men. When it was doing all right they would ease back on subsidy. I think we all knew that they were politicians and they had to be re-elected. They were looking for the right things to happen when the next election came along. I do not think that between 1964 to 1970; that there would have been much difference had there been a Conservative Government.

The Fairfield Experiment [*Fairfield, one of the big six British shipyards, went into receivership in October 1965 when the Bank of Scotland called in the security of a floating charge held on the company. The shipyard, but not its engine works, became the subject of a government rescue particularly on the urging of a Glasgow industrialist, Iain Stewart. After £1 million had been spent on the rescue, a new company, Fairfield (Glasgow) Limited, with a government director on the Board, was formed by February 1966, and the shipyard was made the locus for a new experiment in industrial relations with the declared aim of greater productivity*] was in theory worthwhile, but it was entirely at the wrong time. It was at a time when it was necessary to shrink capacity. Also, if you are going to do an experiment you should choose the most efficient yard, rather than the inefficient one that goes out of business. I think Iain Stewart was very genuine, but lower down there were figures being massaged. You would read about this and that happening, which we had being doing for years. It is just that Fairfield had never done them. They produced a works magazine, and they trumpeted this as a huge advance in labour relations. They were just getting to where we were already. I felt really aggrieved about this. They never achieved better results than were achieved before Fairfield collapsed. They claimed to have doubled or trebled their productivity, but they never gave you the base from which they started their comparisons.

On UCS, by the time that 1968 came, I personally could see only one end to Stephen's and that was closure. We might have gone on for one year, two years, three years, but ultimately it was an inevitable closure. I could see no other way. We were delighted to get into any type of merger that would guarantee our chap's jobs. I personally did not go to UCS because I did not like the set up. It was a question of desperation at that stage. Any solution was acceptable. Rationalisation and government subsidy had to come. I still feel very sore about the way that the government reneged on its promises for UCS. I think that was most unfortunate. They did not support the creditors when UCS went bankrupt. A lot of people lost money. After all, it was seen as a government baby. It very near brought down Alexander Stephen, but for a bit of good work on our side, and the common sense on behalf of the liquidators. The subsequent "work-in" by the trade unions to secure jobs was a very splendid publicity stunt. I had very great admiration for Jimmy Reid [*a shop steward*] and his people. I rather enjoyed it. You knew he was never going to succeed, but it was a wonderful attempt to get publicity. It probably convinced the government that they could not do this and that without consultation.

Interview No. 8: Professor John Rorke, Technical Assistant to the Engineering Director, Stephen, Linthouse, Engineering Director, Denny Brothers, Dumbarton, Chairman, Brown Brothers, Edinburgh, Vice Principal, Heriot Watt University, Edinburgh

At the age of 15, I sat tests which Denny's held for their engineering apprentices and was accepted. I studied at night school at the Royal Technical College in Glasgow, and then full time, gaining a Bachelor of Science degree from London University. I then stayed on as a lecturer at the Royal Technical College until 1950, when I joined Alexander Stephen as Technical Assistant to the Engineering Director. There, I later developed a unit, which became known as a Resonance Charger. It is still manufactured in Newcastle by a subsidiary of Vickers. This seemed to me to be an ideal subject for a PhD so I wrote it up, and completed it in 1957, by which time I had already been invited to rejoin my original company, Denny, as Technical Manager. In time, I became the Engineering Director of the company, and held that post when the firm went into liquidation in 1963. I then came to Edinburgh as the Technical Director of Brown Brothers. There followed a period as Sales Director, and in 1972-1973, I became Managing Director, and eventually, Chairman of the company.

When I think of Denny's the strengths were in their ability to design and to innovate. They were the first commercial operation to build a ship model basin, the Denny Tank in Dumbarton. In this, they were the first company to recognise the scientific side of shipbuilding. From that day onwards the design office in the shipyard was called the Scientific Office. I believe that remained the strength of the company right up to liquidation in 1963. The weaknesses only really became apparent towards the end of the 1950s and early 1960s. It was then that the Japanese influence was taking hold, and it was becoming more and more difficult to win contracts. Denny's had, from mem ory, quite profitable years in 1961 and 1962, but from then on the work just dried up. There were one or two reasons for that. Denny's had traditionally built a number of the cross-channel ferries. I think that there were three of these orders going at the time that we failed to win. Almost certainly, they would have resulted in losses when built by the firms that won the orders. I should say that the Board of Denny had already decided that there was no way we were going to take business at a loss. I think that was crucial in our failure to stay alive.

Denny's, I think, made certain errors of judgement, well, not necessarily errors. We were quite early in the development of hovercraft. Denny decided to go into sidewall hovercraft and were supported on a fifty-fifty basis by the National Research Development Corporation. [*NRDC*] A small craft was designed and built by the Tank staff, and then tested. It gave quite a lot of hope for the future, since it was quite successful. The Denny Board decided the next step was to build one of a five ton lift capacity. NRDC decided that this was too small. They wanted something much larger. Denny said no, it was not practical to take that step. They anticipated that the NRDC would change their minds, but they never did. Denny's built three of these craft without the aid of the NRDC, and that, I think, accounted for the main consumption of working capital. The trials of the craft went ahead quite successfully on the Clyde, but we were foolhardy to put one on the Thames. We learned that the Thames carried thousands of tons of driftwood on its surface. A craft like ours could not survive under these conditions. The company went into liquidation shortly after that period. It is interesting that some of these craft are still operating successfully between Kowloon and Hong Kong.

Another weakness was that Denny's built all types of ships, warships, car ferries, river craft, and cargo ships. Cargo ships were limited in size because Denny's built on the banks of the River Leven. Denny's voluntarily liquidated as a distinct decision, there seemed no hope of operating the company profitably in an age of increasing ship size and increasing competition. When everything was sold, Denny's paid twenty-eight shillings in the pound for every share. The company, therefore, was not insolvent. What it did not have was liquidity at the time.

There were only two members of the Denny family on the Board. One was Edward Denny, who was actually removed as Chairman of the company before liquidation due to the disastrous performance of the company's foundry, Hardie and Gordon, which he was responsible for to the Board. The post of Chairman was taken over by Keith Marshall, whose mother was a member of the Denny family. The other members of the Denny family alive and working for the company were Sir Alistair Denny and his brother, Graham Denny.

The company still had a paternalistic view of things. We really should have cut staff, and closed a number of operations very quickly to save cash, but we did not do it. That must have played a part. In the engineering side of Denny's, and it was common throughout the industry that almost every shipyard had its own Engine Works, we built steam turbines and Sulzer marine diesels. Denny's were, in fact, Sulzer's first licensee. Other shipyards, for example, Stephen at Linthouse, were building steam turbines, Sulzer and Doxford diesels of a range of sizes. There were far too many firms on the engineering side building a variety of machinery. That had to be a grave weakness.

Another came from the shipbuilders who failed to appreciate the savings involved in standardisation. I can remember one particular case at Stephen, where the New Zealand Shipping Company ordered five ships, three from Stephen and two from John Brown. So that was a division already. Each of the ships had an engine with a different number of cylinders, which ran into different types of problems because they were coupled in different ways. The variability was very great. That was quite common. The Japanese adopted a different approach, stating to ship owners what they could offer them, and encouraging owners through price to order six vessels and engines of the same size. It took a long time for these lessons to be absorbed. I think Austin and Pickersgill got it right with the SD14 standard ship. By the time UCS was doing it, it was almost too late.

Looking back on it, the industry lacked a powerful man- a shipbuilder who was a natural leader, and who was highly respected with an ability to order people about. I do not think the industry had that kind of individual [*it did in the 1930s and 1940s in the shape of the Port Glasgow shipbuilder, Sir James Lithgow*]. The only man in that category that I worked for was Len Redshaw of Vickers at Barrow. His shipyard was large by British standards, and he found his niche in building nuclear submarines, and made that a very profitable business, though not necessarily a highly efficient business. The country definitely lacked a strong man. Leadership, in the period I was in the industry, was never there.

In my subsequent career I never found that the management of the British shipbuilding industry was any poorer than that in other industries. I do not think that Denny's were poorly managed, but the style of management belonged to the past. It was not the style that the Japanese were developing. Denny's used PE or PA Consultants in the shipyard, but it was probably too late. A lot of money was spent. They moved ahead, as did Stephen, and put a lot of cash into production and into planning. I introduced these techniques into Denny's during the last few years. I think we were caught by the Japanese. We grew up to believe that the Japanese just copied things. It was not true. The ships were sunk at Pearl Harbour because the Japanese had the finest torpedoes in the world. I do not think that we ever appreciated, until the events of the 1960s and 1970s that the Japanese had skills, which we did not really possess.

We were still running management in the style that had been successful in the previous century. The original families were in the third generation [*some were in their seventh*]. I do not think that you can blame the extant shipbuilding families for the situation. I think the problem maybe started in the prior generation. Perhaps there was not enough done to rationalise the industry during the interwar period [*it did lose one-third of its capacity during the 1930s*]. Productivity was not good. It did not compare to what is now the case in most Engine Works, and we were poorer than the Japanese. Late delivery was a British disease. The Japanese could make deliveries we did not think were possible. I think the Japanese made a bigger effort to get close, to really get close to the management of the ship owners by offering ships at attractive prices, and they were more aggressive in their marketing.

Nationalisation was really a decision of socialism. I think it was wrong. I believe in the market philosophy of Margaret Thatcher, but perhaps not to the extent that she believed that she thought they-she could do no wrong.

Interview No. 9: Fred Walker, Denny Brothers, Dumbarton, Fairfield, Govan, Connell, Scotstoun, and Hall Russell, Aberdeen

At seventeen I was apprenticed to Denny at Dumbarton, and began a course in Naval Architecture at Glasgow University. My career to date has really covered four sections, with each lasting ten to twelve years. My first twelve years were spent as an apprentice, an undergraduate, a trainee in various capacities in the shipyards and ultimately rising to Assistant Shipyard Manager in West Africa, and finally Steel Work Manager at Fairfield, Glasgow. The second twelve years started when I was thirty-one, when I was Shipyard Manager of Hall Russell in Aberdeen, most of the time reporting to the Managing Director who became the Chairman. I then left Hall Russell, which was one of the most profitable yards on a per capita basis in the UK. I came to the National Maritime Museum, Greenwich as Curator of Ships Draughts, and changed my title to Naval Architect. After a decade there, I left to do independent naval architecture consulting work.

Looking back, I was conscious that I was dealing with companies that had something negative about them. Somehow, one never felt on a winning streak in the shipyards. No shipyard seemed to be going in for massive investment, and when one of them did, it never seemed to work. It was a very sad situation. From the mid 1960s there was a tremendous demoralisation amongst the group that I was in, the men of thirty and upwards. The loss of senior men and graduates at that stage was absolutely criminal and critical.

If I can digress for a minute, I want to go back to when I was an apprentice at Denny's. It was the posh place to train, the public school of the British shipbuilding industry. They trained at one stage more undergraduates than the rest of the Clyde put together. Even today when you go to London you find more Denny men than of any other background. During that period it was made very clear to you that you were there on sufferance. The company made you serve your apprenticeship of five years, and allowed you days off to go to university. By contrast the attitude of the apprentices was that if we were undergraduates we might, as a matter of decency, go to the shipyard and give them a summer's effort. We were poorly paid. The first three years were horrible. The attitude of the men and staff was that they had seen students come and go; it was totally negative. There was a great anti-education lobby in what was one of our great shipyards. What it must have been like to have been a university apprentice in a place like Blythswood [*the Scotstoun tanker specialists*] hardly bears thinking about.

Moving to the 1960s and being in a group of younger guys of executive rank, it has to be said that we all felt that we were going nowhere. They were not being paid very much. There were no perks. Shipbuilding was not seen, as it once had, as being a place of social consequence to work in. Now that may or may not be important, but if you were going to keep young executives in, they had to feel that their jobs were worthwhile, and that they were going somewhere.

One of the difficulties, I think, in the period of the 1960s on the Clyde, was that many of the yards were family owned. If you were not connected to the family you were going to have a more difficult path to get to the top that if you were connected. I am not suggesting that the Denny's or Stephen's were not good about it, but the fact was that they had a vested interest in the shareholding, the control and the directorate within their own yards. It you look at most of the big yards that were family-owned, Connell's was a good example, they did have outside directors on the Boards, but they were very much manipulated by the Chairman and the Managing Director. In the case of Connell's both of these positions were held by family members. They were quite prepared to pay high salaries for general managers and people of that sort who sat on the Board, but it was clear that decisions were being made by the family and the trustees of family investment trusts. I think that this demoralised the young crowd, and I, myself, for example, made two moves. At the end of my apprenticeship I had not finished my degree. I decided to go to another yard to gain further experience, and to go abroad, which was not regarded as good form on the Clyde.

Being an undergraduate, I went out to work at Burmeister and Wain in Copenhagen. When I returned, I offered my services to Denny. By that time, however, they were on the way down, and they wisely asked me to go elsewhere, and so I joined the Lithgow Group [*in the 1950s, the largest private shipbuilders in the world*]. I found myself spending four years with the Fairfield Shipbuilding and Engineering Company at Govan. I would regard the old Fairfield Company as about as arrogant a bunch of shipbuilders as you could find. They did not seem to make money. They were heavily over-manned. Their managers had a strange lifestyle, dressing in Bowler Hats. In many cases, I felt that the directors were not in touch. As the years went by I became friendly with the Managing Director, Jim Lenaghan, and his Deputy, Derek Kimber, and thought highly of them. I think Derek, intellectually, was a giant of a man. Somehow or other, it was a Board that was very remote from the shipyard, and yet the Managing Director would sign for everything.

During this period, people were leaving the company. I ultimately left and joined Connell's as assistant shipyard manager, where I spent eighteen months. It almost drove me to a mental home. I was fired, publicly, once, in front of six hundred men for carrying out what was later shown to be a correct decision. After a big enquiry, I was taken back the next day without an apology and was told that I was on one month's probation. That was the nearest that they would get to admitting a mistake. I was only to glad to leave there and went out to Africa for a short period.

I returned to the Clyde, and the next eighteen months were fantastic. It was the time of the Fairfield Experiment, and I joined as a planning engineer initially, at an inflated salary. Shortly after arriving, I became Steelwork Manager with a team of around eight steelwork managers. I worked for most of the period heading a team of about four hundred, foremen and men, building bulk carriers. I worked very closely with the head foreman caulker, Eddie Docherty, who was later made up to a manager. We were a little bit suspicious of each other, but got on well later on. I began to realise that the foremen were very fallible, although their hearts were in the right place. Very few were lettered people in any sense, and they were overawed by managers who wore the black and gold tie of Glasgow University. If they knew you could handle a ship, and you could handle the men you were in with them, because they knew, technically, they could not argue with you. I think that would be my first sadness in looking back, that we lost that technical professionalism which existed in the industry.

The Fairfield Experiment was badly needed. At the very beginning both the men and the managers welcomed it, and the atmosphere was unbelievably good. If you had said to the men, I want you to strip down to your underpants and dance barefoot on the frosty decks because it would help the shipyard, they would have done it. We, as managers, were in before the men, and we were told to say good morning and smile. The atmosphere literally changed. Sadly, a lot of shop stewards developed powers that were quite extraordinary. Possibly, I was emancipated, but not emancipated enough not to have to wait outside the ship stewards convenor's office awaiting his convenience to see me to discuss a matter. It was also upsetting when you went into the office and little respect was shown.

I thought that the Geddes Report was great. Looking back on it, I think that the groupings of yards could have been done differently. I think one or two yards were grouped together which should not have been. Again, it has to be said, that it was a first attempt. If government and industry had not been in such a panic themselves, they might have made a better job of it. One aspect that became very clear was that money was being sunk into the shipyards to keep them going. Money should not have been put in that way, but rather as seed corn. I am not certain how it could have been done. People go on about Lame Ducks, and I have to admit that I do have some sympathy for the lame duck argument. The money should have been used for re-development and training. These were two areas which we signally failed to do anything about.

I think that in all the yards planning was done by one or two people at the most, and the final decision was taken, inevitably, by the Managing Director. Many of them had the very bad habit of not consulting the people who were around them. There was a certain need to keep the populace in ignorance, I think. You were very rarely made party to what was going on. Boards wanted to rubber stamp what the Managing Director and Chairman had already decided. You never really saw the financial figures for the shipyards. At Hall

Russell, which was part of a merchant banker's, none of us was ever able to find out exactly how much money the company was really making.

The planning departments were there, on the whole, for supplying the managers with detailed planning on a day to day basis, and for planning production for possibly two years ahead. Very few of them were used in terms of strategic planning. There was no such thing as a strategic planning department. In Fairfield, I remember being concerned that we were not picking up Scandinavian orders in Scotland. I wrote a memo on the intelligence function in a shipyard and put it to the Chief Executive. I argued that we were working a great deal in the shipyard with method analysis, with quality control, and with every souped-up organisational method, but, in actual fact, we were not sorting out the problems with our customers. We did not really know what our customers wanted. If they did not want 70,000 ton tankers there was nothing much we could do about it. There were some ships that we did not want to build at Fairfield, but, in general, we would take on any ship. I felt that there should have been a market analysis first. I was told that, in due course, I would be moved out of the shipyard and made understudy to the Chief Estimator. It never happened though.

If I was forced to give a blanket answer on whether British ship owners let the industry down, my answer would have to be yes. On nationalisation, I am not certain that British Shipbuilders was the answer. Marketing turned out to be poor, and many of the appointments made at the top level were very sad. It was utterly stupid to have a Head Office in Newcastle on Tyne. It created all sorts of problems, and to the shop floor there seemed to be too many executives floating around. In my opinion there should have been a lot more toughness in the way that it was operated. I was one of the people who got on with Graham Day, when he was Chief Executive Designate. Without planning it, I met him along with several directors of Hall Russell. I told him, on behalf of the managers, that we had several positive messages, one being that we should retain our identity. Looking back, I am not sure whether retaining our distinct shipyard identities was the best thing after all.

British Shipbuilders did one clever thing. They were told to shed yards, so they shed Yarrow's and Vickers. I have always thought that the naval yards were not the lead yards. They were faking a profit, and they should have been sold when they were fashionable. But the high technology should not have been sold. That should have stayed with NESL, and Govan Shipbuilders [*the successor company to UCS*]. They [*the government*] should have stayed with NESL and with Govan. I went to a meeting of the Parliamentary Maritime Group addressed by a Government Minister, Tom King. He said that one or two exciting things had happened like the sale of Harland and Wolff at Belfast [*the only large firm which was not nationalized as a part of British Shipbuilders*]. Now, for the first time, he said, Britain can show the world how to build ships. I could not believe my ears! In reply I told him that his government had saw fit to sell Govan [*a yard which had been subject to massive state expenditure, and was thoroughly modern, particularly in steel work facilities*] to the Norwegian Kvaerner Group, and to close down NESL. King replied on the latter, that they had to sell NESL because that was the deal we made with the EEC regarding Enterprise Zone status for Sunderland.

This lack of understanding has been symptomatic of the industry.

Interview No. 10: Jolyon Slogget, Naval Architect, Denny Brothers, Naval Architect, Director of Finance and Development, Houlder Brothers, London, Executive Deputy Chairman, Houlder Offshore, Managing Director Marketing of Merchant Ships and Offshore Fabrication, British Shipbuilders

I started as an apprentice ship draughtsman at Denny, and was released each winter to go to Glasgow University to the sandwich course in engineering, specialising in naval architecture. I graduated in April 1955, and completed my apprenticeship in October 1956. I then worked in the design office for six months before undertaking National Service in the Navy. I returned to Denny and worked a further fifteen months there until I joined a British shipping company of medium size, operating about thirty ships, Houlder Brothers in London, as an assistant naval architect. In 1965, I became Naval Architect, and in 1968, manager of new projects. Two years later I became personal assistant to the Chairman, and in 1972 joined the Board as Director of Finance and Development, and retained that position until 1978, when I left to join British Shipbuilders on a three-year contract working on offshore enquiries. I left in 1981 to do consultancy work, mostly in the offshore industry and in 1986 became Secretary of the Institute of Marine Engineers.

Denny's were still a family business when I joined them, and it was a small friendly yard. The benefit of working there, from my point of view, was that they built a wide variety of ship types. They did not build very large ships because they had a size restriction due to the River Leven. I remember at that time, that Edward Denny, when he was Chairman, complained bitterly about the time lost through demarcation problems caused by the various unions. In those days this was probably the biggest single problem the management had to face. Denny believed that thirty percent of time wasted was due to restrictive practices. When, at Denny's they started a Planning Department, this produced a number of ribald comments, particularly from the senior foremen in the yard.

Demarcation was obviously a very difficult problem. Men still remembered the 1930s when they had been on the Dole [*state unemployment and social security benefits*] for long periods. In the 1960s men were still reporting to the yard on a Monday morning [*casualization of employment*], and the yard would take on the number of joiners, carpenters or welders it wanted, and the rest were turned away. I remember in the mid-1960s the Tyne shipbuilders used to get together to compare the number of men that they were going to want. When the men are on a market like that, you have to expect them to behave accordingly.

The shipbuilding industry started to contract in the early 1960s. William Gray at Hartlepool closed, and Denny's closed in 1963. Government due to political pressure from Members of Parliament in shipbuilding constituencies began to become involved, and this later led to groupings and Government takeovers of firms in difficulties. In the run up to nationalisation the resignation of Graham Day from the Organising Committee of British Shipbuilders was unfortunate, had he been in charge, things might have been different. The other unfortunate thing was that a large part of the ship repair industry was nationalised at the same time. A number of ship repair companies were losing enormous amounts of money, and should have been shut down straight away. I recall at one meeting it was suggested by the Board Member for Finance that a ship repair yard on the Thames should be closed because of huge losses. The Chief Executive, Casey, said that we could not do that because it was in a Government Minister's, Peter Shore's constituency. Had the ship repair yards not been included in nationalisation then the drain on resources would have been lessened.

I think the main difficulty with British Shipbuilders during the time I was there was the independence of the shipyards. I saw it initially from a marketing point of view where, because of the Act, it was very difficult to get a co-ordinated sales effort. Not only that, the managing directors and chairmen of the various shipyards had been used to running their own show, and they resented having a British Shipbuilders Board over their heads to whom they had to report, and who could give them directions. Towards the end of my period, I was trying to sell drilling rigs to a major international organisation, and they sent somebody to visit three shipyards that were possible builders. When the director came back he told me that my major problem was the barons in the shipyards.

One important factor was the introduction of the Intervention Fund. That was vital. British Shipbuilders played an important role in dealing with it alongside government. Had all individual yards been dealing with government on it, I think it would have been totally chaotic. Doing it all through British Shipbuilders meant that it was done on a more rational basis. One major error was that there was no general policy set down as to the sort of ships that British Shipbuilders ought to be trying to build. Unfortunately, in the shipyards if there was a hint of a bulk carrier order, a nice easy ship to build, then all the yards management would be rushing out to get this order, however nebulous it might be, rather than try to develop some more complicated vessel. For example, cross channel ferries, which we used to build, were the sort of high technology vessels that we ought to have been concentrating on. But all management would do was to take the easy way out. There was no direction [*in the early years*] to push them into areas high technology. They would rather run a mile than get involved in that area, even although that held out hopes of work to far greater extent than competing with the Japanese over the construction of steel boxes [*tankers*].

When a typical enquiry for a vessel came in we had to send it to five or six yards. The big problem was that as a nationalised industry we got an awful lot of enquiries, and because it was known that we were short of work, everyone expected that every enquiry put forward would have the red carpet treatment. So, whereas a private shipbuilder could simply say that they were not interested in an enquiry, we could not do that in British Shipbuilders. There were one or two cases when we knew that an enquiry was a no-hoper, and accordingly was not dealt with. As a result, there were even questions asked in the House of Commons. Individual yards did get enquiries, and were supposed to notify Headquarters, but, of course, they tended to be a bit slow in doing that. That also applied to offshore enquiries. Scott Lithgow felt that they should get all offshore enquiries, but my instructions were very clear, that I was to send these enquiries to everyone who I felt was capable of answering them. British Shipbuilders acted as a central clearing house for enquiries, and once distributed, individual yards would deal with owners.

There was some truth in the charge that British Shipbuilders was over bureaucratic, and that some officials had no prior knowledge of shipbuilding. That goes back to the very top, in that the Chief Executive, Michael Casey, as far as I know, had never run an organisation of any sort previously. I can not understand why a man like that was ever appointed. I think Admiral Griffin, as Chairman, has been under-estimated. It was a very difficult situation, and a silly one, in that you had an executive chairman and a chief executive and I think the fact that they appeared to work together was probably due to Admiral Griffin's charm. It was a difficult situation and he achieved quite a lot. If the right man had come in as chief executive then things might have happened very differently. Griffin had originally said that Headquarters would have no more than eighty people in it, but it developed quite out of control in my opinion. People who are not fully occupied will find reasons to expand their departments, and that is what happened at British Shipbuilders.

Having two headquarters was political. It was forced on British Shipbuilders by the political masters. The result was that the London headquarters was really where all the real work was done, because we were working very close to the Government, and had to be. British Coal, British Steel and all other nationalised industries headquarters were in London. The result was that an enormous amount of time was spent by senior executives travelling from Newcastle to London for meetings.

Another political interference was the Polish order, which was an absolute disaster as for political reasons it was announced before it was signed.

The result was that the negotiations were then a farce because British Shipbuilders were negotiating with both hands tied behind their backs. The Poles just dictated the terms. It was a bad bargain as the ships were under-priced [*but did provide yards with much needed work when no other merchant orders were forthcoming*].

The warship yards were profitable, and the merchant yards tended to unprofitable. Indeed, one or two were sailing very close to the wind before nationalisation actually saved them. For that reason, I think the warship yards resented it. I got some insight into this as I was on the Board at Vickers. It was illogical in a sense to privatise the warship yards, first, because they were on Government work all the time, but on the other hand, you can not privatise a loss making organisation [*you could, and the government did in the case of Scott Lithgow, the first privatization*]. I suppose that if you are going to privatise then you privatise profitable areas first.

Later on, what happened to British Shipbuilders depends to some extent to your political views. If the industry is going to survive in the long-term it has to be profitable. No Government is going to be prepared to pour money into anything indefinitely. Possibly the spur of privatisation was that it forced profitability into those areas which could be made profitable. I think it was a tragedy that NESL at Sunderland was closed, when all the indications were that demand on the merchant side was picking up. That is the one merchant yard which should have been saved.

I do not think that anyone thirteen or fourteen years ago would have expected to see the decimation of the industry. The knife should have been taken out sooner and the funds should have been concentrated on the more profitable areas from an earlier stage.

Interview No. 11: Dr. John Brown, John Brown and Company, Clydebank

I started at Clydebank as an apprentice in the shipyard drawing office in 1919, and spent several years in the general drawing office. Having graduated from the University of Glasgow in 1923, I was then transferred to the ship design office as a naval architect, and continued there until 1927, when I was seconded to an associate firm in Spain under the direction of a former Clydebank man. I had two years in Spain as a designer, and as an estimator, coaching Spaniards. I returned to Clydebank where I was appointed Assistant Naval Architect. I served in that capacity from 1929 to 1949, which was a long haul. In 1949, I was made Naval Architect and Technical Manager in charge of the design and drawing office. From 1949 to 1959, I occupied that position, and then I was made Managing Director in succession to Sir James McNeil. I held that position until the end of 1963, when I handed over to John Rannie, with the idea that he would handle the construction of the Queen Elizabeth II. I had actually supervised the design in conjunction with Cunard, and supervised the tender which was eventually accepted. Since I was approaching retirement age, I was made vice Chairman, with a consultancy role, and stepped out altogether in 1967.

Our firm, I think, was in an individual position because of our size. We were, if not the biggest, one of the biggest in the industry, and in many ways that was strength. It meant that our main products were the very large ships, the passenger ships, and although we did compete in the market for smaller ships, our weakness might be that if there were any failure in the market for big ships we would be the sufferers. Another aspect in the situation of the firm; we were the main employer in the Clydebank community alongside the Singer Sewing Machine Company, and so there was a fair amount of folklore around the shipyards. The people of the district spoke of the yard as something that had always been there, and would always be there. There was a great deal of father and son associations. That particular aspect of things could well have been a weakness. Fathers who had grown up in earlier days when conditions were admittedly poor in the shipyards, passed on their knowledge, their experiences to their sons, and I think that engendered an attitude, not mistrust, perhaps, but a feeling of antagonism, a "them and us" attitude, if you like.

Over the years we did take on intermediate sized ships, and we were also a warship building yard. For the big passenger ships, we were in competition with Fairfield, Harland and Wolff, Swan Hunter and Cammell Laird. In warship building, we were in competition with the same firms. These were the big firms. Over the years, as I have thought about it, I came to the conclusion that our management system was weak. We had a London Board under which we operated for a long time. We had a fair amount, a full amount, of local autonomy. In 1953, the Clydebank establishment was hived off as a separate company, and then we had complete autonomy, although the Group Chairman and his vice Chairman were on the Clydebank Board, but they did not exercise detailed management control. I do not think this was true of most of the yards, but in our yard the local board was made up of graduate engineers and shipbuilders, skilled technically, but in my more mature consideration, not skilled in management or finance.

We had a Secretary, the only one on the Board who was not a graduate engineer, and he was responsible for handling the books, but although he was an able Chartered Accountant, he did not come into policy making. On the management side; and I personally admit this, I never was very happy about people going off to management courses. I regarded them with a fair amount of suspicion. I still do not know whether I was justified or not. People grew up as practical men in the shipyards, and in our case with a thorough technical training, but the people who handled the labour relations, and the actual construction work outside the offices, in most yards, would come up through the management side of things. I do not know if, at that time, management training, in the modern sense, was available. I think it was only beginning to come in, and we were all suspicious of it. I think we were right at the time, but I think today we would change our ideas altogether.

We did not research the market as seems to be the modern practice. I think most firms, but Clydebank in particular, was fortunate over the years with having customers with which we associated regularly. It was very noticeable in the case of Clydebank, because, apart from Cunard, we had an association with the New Zealand Shipping Company, which was very profitable over the years, and probably led to a degree of complacency. You did not have to go out and seek orders. The skills needed for the larger passenger ships in a way blended with the skills needed for warships, so that the two sides of the market were an advantage to Clydebank. When things became slack, and eventually some of our customers, looking to their own interests, went elsewhere, we had to think about it. We had no personnel doing market research. Seeking orders devolved on the Managing Director, and I kept in contact with some of our former clients, and indeed other people in London to find out if any possible future building was likely, and seeking the opportunity to go for new clients. Everybody else was doing that too. I did travel abroad. I was twice in Moscow. Our London side had a connection. The second time I went there I went with someone from Scott Lithgow and James Stephen. There was a possibility of building fish factory ships, but it came to nothing. I visited Gothenburg and Stockholm. On these occasions we took no preconceived proposals. The marketing side of it was difficult.

Did British ship owners let shipbuilders down, they could not help themselves, and they did. But they were as hard up against it as we were. Our clients in New Zealand, we had built about twenty ships for, and had very close friendly relations, when it came to the bit, they went to Japan. At that stage it was not the technical side of it, it was cost. We generally believed that we were subject to unfair foreign competition.

On labour, our Shipyard Directors, as we called them, handled labour matters. At Clydebank we had a multiplicity of unions. We hear about restrictive practices, and no doubt they existed, but it seems to me a more important problem was this one of numerous unions. You did not have very often a general strike in the yard, but you had one particular trade going on local short term strike, which was very upsetting to the production programme. When we first introduced arc welding, and a welding machine was produced to do a straight line weld, the welders demanded that the machine required two men. It was a straightforward one man job.

During the 1920s the management was very harsh, but under John Rannie's regime, less so. Rannie cultivated the workforce, but I think at the expense of concessions. When a problem came up on the river as a whole, and the shipyard managements got together on it, there were always the two opinions. Somebody would say let us face a strike, and others were not prepared to do that. In our case we usually conceded, rather than face disruption. We never had a personnel manager, which may seem strange in modern terms, and I think the only place in my time that had one was Fairfield during the Experiment. We were of course a composite firm with our Engine Works. They had a personnel manager, but he had limited responsibility.

I wondered often if we were the victims of the Geddes Report. We were persuaded, coerced if you like, into going into UCS. I sometimes wonder if the Clydebank yard could not have continued to operate on its own. When we were incorporated into UCS, there is the tale that our chairman [*Lord Aberconway*] offered them the yard for £1. That's folklore too. [*It was, but some years earlier, Lord Aberconway offered to sell John Brown's for £1 if the government would finance a scheme to build giant tankers by straightening out a bend in the river at Newshot Isle to enable giant building docks to be constructed*]. At that time we were looking for support from the Government Minister, Tony Benn, which he failed to give. I do not think that we got the support once they had created UCS. [*the Labour government pumped £20 million into UCS from 1968 to 1970*].

The Fairfield Experiment was unfortunate. I think it had the seeds of what the industry needed in management procedures. Iain Stewart gave a famous lecture, and I think it was sound thinking. We were not ready for it at that time. The people that he brought in to Fairfield were not shipbuilders. In fact the man he depended most on came from Singer Sewing Machines in Clydebank [*Jim Houston*]. He was a modern thinker, with possibly modern ideas ahead of his time about management. Had he been five years later it might have worked. It was never really responsible there. It was not a happy marriage, but the ideas were sound enough.

There is an aspect about the UCS work-in of which I feel annoyed. The famous Jimmy Reid, whom I knew, claims that he led the work-in, which saved the yard. I think that the correct version of that is that Robert Smith, the Liquidator, doing his best for the creditors, had to get the ships which were in the yard finished, and therefore needed a labour force, so he was going to employ the labour force and pay them. No matter whether they claim that they walked in and kept the ships going; in fact, he was employing them and saving the ships for the creditors. So the famous myth about the famous work-in is nonsense.

Interview No. 12: Graham Strachan, Alexander Stephen, John Brown, and Scott Lithgow

I started an engineering apprenticeship the day after I left school in 1950 at Stephen's of Linthouse. After two years, I went to Glasgow University for my first year in engineering. I then went to Cambridge and completed a five-year apprenticeship, combined with a university degree course in 1955. After National Service in the Royal Navy, I obtained a job in the design office of John Brown's at Clydebank. In 1963, I was appointed Technical Director, and shortly afterwards I became Engineering Director at the age of thirty-one. I remained there until 1966, when, following the recommendations of the Geddes Report, engineering was separated from shipbuilding. I initially became Director and General Manager, and latterly Managing Director of John Brown Engineering. I left in 1984 to take up a position with Trafalgar House, when they acquired the Scott Lithgow yard.

When I came back from my National Service in 1957, I remember that the order books at John Brown stretched well into the early 1960s. John Brown at any one time could be building a large passenger ship, a couple of tankers, a couple of bulk carriers and so on. They had an incredible mix of products. That was seen as strength by some and a weakness by others, because one could say that had they concentrated on one type of vessel they could have done a lot better, rather than dispersing their activities and resources over a range of products. One of the weaknesses arising from this was that everything was made to the highest quality demanded by the one type, whether that was a passenger ship or a naval ship. So a bulk carrier was constructed with too high a finish than was absolutely necessary. The problem was that the owners expected a very high standard of finish from a yard like John Brown, whereas they would have accepted a lower standard from the Lower Clyde, which only made this type of ship, and did not build passenger ships. [*Historically, the Lower Clyde yards did build passenger ships, and the district did not concentrate on one type of ship*]. It was impossible for the workforce to work on dual standard of quality. They tended to work to the highest, and the ship owner insisted on the John Brown standard.

On foreign competition, the industry was well aware of the threat from Japan and certain European countries from the late 1950s. I think that the UK was holding its share of the market reasonably well. The world market was expanding, and that increase was being picked up by others. Various delegations were sent to Japan to see how they were doing it. I remember two of our directors coming back and feeling very depressed with the advantages Japan had, particularly with Greenfield sites, money being made available, and wondering how we would compete. My recollection at the time was that substantial investment was being made in John Brown and in Fairfield, which was principally aimed at the tanker market. The consensus was that the size of tankers would increase. We reconstructed our West yard with two big berths capable of building 80,000 to 90,000 deadweight ton tankers, but no sooner had this been done than we were talking about 100,000 to 150,000 deadweight ton tankers. It was one of these things when everyone was caught out. Nobody realised how quickly the scale of tankers would increase.

It's fairly well known that marketing, as we know it today, was almost non existent in the shipyards. It was done by the Managing Director, or by his senior colleagues on the old boy network basis. One tended to concentrate on the old established customers, even to the extent that when the yards were busy other customers were turned away. Market research was what you thought you would be building in five years. Up until 1966, John Brown never had a sales department as such, although we had one on the engineering side.

The problem with shipyards in those days was that you had half a dozen extremely talented people on your Board of Directors, who had probably been steeped in shipbuilding all of their lives. Most, if not all of them, had a degree of some sort. Below that your manager tended to be someone who had come up from the shop floor, excellent at his job and in his knowledge of the industry, but very weak in organisation and methods because he had never had any training to do this. It was almost unknown to have a graduate as a manager in the company, unless, perhaps, it was your chief naval architect. So you have this middle group with very little tertiary training.

Policy was pretty *ad hoc*. The naval side of the business provided a steady level of profits, because it involved essentially reimbursable costs plus contracts. That gave a certain base load, but was not enough to keep the yards busy, unless it was Yarrow's, which was essentially a warship yard. The passenger ships would provide the work content, which would last three or four years and employ several thousand people, and hopefully, at the end of the day, produce a good profit. It took up a lot of resources, and did not optimise the steel working facilities, so you needed tankers and bulk carriers which were big on steel; and less so, on the outfitting side. There was always this tussle between the priorities. Our policy was to keep our options open for better or for worse. There was continual debate about what would happen if we gave up building passenger ships or tankers. We were frightened that if we decided to cut out something, it might be the wrong decision. The one thing John Brown did get right was jack-up oil drilling rigs. We were the first yard in the UK to build them, and built three between 1964 and 1966. That was an example of adapting to a new situation, but still building frigates, passenger ships and tankers at the same time.

Geddes analysis of the situation was very good. Not too many people would disagree with the bulk of what he said, and most of his conclusions. The carrot was the financial incentive to group. UCS came out of Geddes. My recollection was that they had good objectives in that they started to centralise marketing etc. They had a corporate headquarters of new people who were brought into the industry, some of whom had no previous experience, like Tony Hepper. They also brought in consultants who thought they could sort the industry out, but found that they had bitten off more than they could chew. Another layer of overheads was created. The rationalisation did not achieve the cost cutting that was envisaged by Geddes. Everyone kept their own drawing offices and so on. It turned out to be desperately under-capitalised. They fell into the trap of assuming, when they were bidding for work that productivity improvements would be made. They cut the man hours and therefore the costs. The problem was that they were not achieved, so they quickly ran out of money. In 1970 or thereabouts they went into liquidation. It was the right idea, but it was carried out the wrong way. They were not ruthless enough in cutting back and in centralising. The work-in at Clydebank, in particular, did show what could be done to change the policy and attitude of the government, and the result was that the Clydebank shipyard was not closed, but was sold to the Americans [*Marathon*] for construction of drilling rigs, which was the next best thing.

I never particularly favoured the Fairfield Experiment. I did not like the people who were brought in to run it. They were theoreticians rather than practical people. You could say that it was worth trying something, than to do nothing at all, but we will never know whether they succeeded or not, because the whole thing was stopped in the middle. Up to the mid 1960s the industry had the worst record of lost time through disputes except perhaps the motor car industry. Most of these were as a result of wage demands, either comparability between different unions, but more particularly skill shortages, particularly among steel workers. If you needed another hundred steel workers, the only way to get them was to pay sixpence more an hour and pinch them from Fairfield. Then, your joiners would want to be made up to that rate. These disputes over differentials no doubt led to this terrible record of lost time due to disputes. At John Brown there was hardly ever a demarcation dispute, simply because demarcation as it existed was well defined and understood by those who had to operate it. It was mostly wage disputes, and they happened on a regular basis. They should have bought out demarcation and introduced flexibility and interchangeability between trades.

On government, Labour looked more favourably on shipbuilding that did the Conservative Party, but neither did anything until it was too late. What really had to be done was to cut back dramatically on UK shipbuilding capacity. On nationalisation, I doubt whether many firms would have survived financially but for it. Money was more freely available, but it was misspent. Instead of it all going into investment much of it went to subvent losses. You got a rather sick industry continuing to be sick because they always knew that the Government would make up the losses. You only have to look at the figures of the hundreds of millions of pounds that the shipbuilding industry lost over that period. That must condemn nationalisation for a start. I suppose there must have been some good in that we still have a shipbuilding industry because of it. Investment was made in some yards, which put them on a par with some of the better yards around the world, but it was probably at a time when the shipbuilding world market was weak, and they could not command the prices to make profits.

Trafalgar House was the first company to buy out a nationalised shipyard in March 1984. It was a very bold decision to take over Scott Lithgow, although it was more or less handed to them on a plate. They were lumbered with a very difficult contract [*a highly sophisticated semi-submersible, dynamically positioned drilling rig for Ben Odeco, subsequently named, Ocean Alliance*] which lost Trafalgar House an immense amount of money [*the company eventually won damages from British Shipbuilders*]. They bought the yard essentially to complete their repertoire in the offshore field. They thought that they would be a forthcoming and continuing market in semi-submersible drilling rigs. No sooner had they bought Scott Lithgow than oil prices collapsed, and there was a surplus of rigs on the market. Very few have been built since anywhere in the world. An additional factor was that being a privatised yard designated by the Government as an offshore yard, Scott Lithgow was not eligible [*from 1980*] for European Shipbuilding Intervention Fund subsidies [*a subvention to allow European merchant shipbuilders to tender on equal terms with Japanese and Korean yards*] like Swan Hunter, and other warship yards [*this is wrong: warship builders under the Treaty of Rome were not allowed to receive Shipbuilding Intervention Fund monies*].

The thing that always impressed me about being in a shipyard and being on the engineering side of shipbuilding was the skill of the workforce. In those days, before automation, it was incredible what they could turn their hand to. No task was too difficult for them. One can criticise them for not being motivated, which was probably true most of the time, but if it came to a real crisis they would respond magnificently. The turbines [*designed by Pametrada, and which had broken down*] on the *QE2* in January 1969, were re-designed, rebuilt with new blades, and delivered back to Southampton by the 8th of March. I doubt whether there was anywhere else in the world that could have achieved that in the time it took. The ship went back into service in April. That's just an illustration of the skill of the men, and the way in which they could respond to a crisis. If only they could have been motivated to do this 365 days of the year, perhaps we would not have this problem.

I think it was inevitable that British shipbuilding declined from the point of view of overseas yards starting up with Greenfield sites with American financial assistance in Japan, and the same in Germany. Initially the substantially lower wage rates paid in the Far East meant that there was no way this country could compete on price, certainly through to the late 1960s and early 1970s. When the wage rates began to come more into line with our own, their attitude, to the work and their productivity by that time was far superior to ours. In some ways it is a bit surprising that we have any shipbuilding industry left whatsoever.

Interview No. 13: J.F. Starks, Royal Dockyards, John Brown and UCS

I started at the age of fifteen as a shipwright apprentice at Devonport Dockyard, and managed to get a Royal Scholarship in Engineering at London University. At nineteen, I became a Constructor Sub-Lieutenant, and did three years at the Naval College in Greenwich getting a first class Certificate in Naval Architecture. When the war began, I was sent to the Admiralty establishment at Bath, as an Assistant Constructor in the Submarine Design Office, and I attended all the sea trials of submarines built at Vickers. Before the war was over I was appointed a Constructor Commander on the staff of the Royal Naval Mission in Washington DC, working with the US Navy. After three and a half years there I returned to Bath, where I became a Chief Constructor. After five years, John Brown's asked me to become a director, and I eventually joined them in 1960, after the launch of the Dreadnought [Britain's first nuclear-powered submarine] as Technical Director and Chief Naval Architect. I spent from 1960 to 1972 with Brown's and latterly with UCS with whom I was a Technical Director. Prior to that, I had been made Deputy Managing Director at Clydebank. When UCS hit the rocks, at the age of fifty-five I joined Vosper Thornycroft in Southampton to become their representative in Rio de Janeiro for five years before I retired.

The primary strength of John Brown was our reputation, fully justified, as builders of large passenger ships. It was also a good firm in that it could build anything. We had also a very good design team and a fine Joiner's Shop. We had very good relations with the owners, and I must emphasise that our reputation was enhanced by the *Queen Mary* and the *Queen Elizabeth*, which were tremendous assets during the war as troop carriers. The standard of workmanship was high. I think our prices were high, but so was our quality. For instance, the Royal Yacht, *Britannia* was built at Clydebank.

To come to the weaknesses; I think there were too few top men. There was the Managing Director, the Shipyard Director, Engineering Director, Technical Director and a Finance Director. There were only five of us and we worked too hard. We tried to do too much. I do not think that we were good at cost control. That is a wide subject, but it is necessary to appreciate that after the war, in the 1950s, orders were easy to come by. The prices were good, and there was a fixed fee system or cost plus system. They were not used to changes, and the system did not have tight cost control built into it. They were awful slack in charging. Nor was it unusual for work to be charged for a different ship, than the one for which it was used, a practice not unknown elsewhere, and certainly one which had been in existence for decades.

When passenger ship orders disappeared, we were stuck, and did not switch fast enough. The Managing Director and I toured the oil companies' offices in London looking for some work out of North Sea Oil. We found one company, who persuaded us to build an oil rig (*a relatively unsophisticated jack-up rig particularly suitable for work in the Gulf of Mexico*]. On the first one, we made a profit chiefly because the owners were very helpful to us. But it was not the best type of work for our yard. Earlier on, the Board was horrified when I suggested we abandon the building of passenger ships due to the age of the airliner, and concentrate more on warships.

The whole of the industry suffered from a lack of investment. Brown's did at least build a brand new Drawing Office, and also invested large sums of money in the Engine Works. On management, we could have done better. We did not have specific management training, although this is easy to say with hindsight. If you had a good man doing a job, you were happy with that situation and left him to it. On labour relations we were average, but the industry was bad. We did not have sufficient modern planning, and there is no doubt that we did not always keep to delivery dates.

Later, there was a tendency for people to decry the quality of Japanese built ships. Their ships were first class. I found in Brazil, that you could take men who were almost out of the jungle, and who had never seen a ship, and give them a little test on welding. If they showed any adaptation at all, you could train them as a welder. They would do exactly as you told them, and you would have a crack free weld. If you tried to do this with a man already skilled in Britain, he would have his own ideas. In foreign yards they were able to cut out all the bad practices. If you went to Japan you could see the welding of a hull of a tanker with a machine run by one man. Over here, you would have a man, a mate and an apprentice. You would give in on manning levels, or come to some compromise. I also think that the nations abroad worked harder, because they had been defeated, whereas in Britain, after the war, it was easy going. We had, in my view, over-employment.

Going back to labour relations, some of the Draughtsmen were the worst. They tended to pick out one firm at a time, and realised that they could wreck progress on ships. If the orderings and drawings were not done, then nothing could move. They did not, in many cases, have the nerve to go on strike, but carried on go-slow tactics, and did silly things like not answering the telephone. Worst of all they operated a closed shop, [*100 percent union membership*] and would not work with non-union personnel. Demarcation was bad, and the problem was that the men did not see the seriousness of it. What we should have done was shut the place. I will give one example; in 1965, I think it was the plumbers on this occasion. We were building a liner, which we had taken on at a low price, and it became clear that we were going to lose on it. The Shipyard Director and I called a beer and sandwiches meeting on a Sunday. We got all the plumbers in and showed them the figures. We gave them all the reasons why we had to do better, and they were sympathetic. When I arrived home that night, I believed that I had made progress, and that the men realised how bad the situation was. They went on strike the next morning. The stewards must have went back to the men and told them that they had got us over a barrel. We have got to give in. This was the attitude. At that stage, we should have shut the yard, by way of a lock-out until the men came back to work. The industry tried a lock-out of the Draughtsmen once, and they went of on holiday. I received a postcard from Blackpool saying "wish you were here."

I thought international competition was generally unfair. Sometimes there were prices quoted that were impossible. It was when you saw a big difference on the price of materials in tender prices that you began to suspect governments. We should have reduced the number of shipyards. With hindsight, it was no good, to my mind, to bring together, as in UCS, a number of loss making yards. How could they make a profit without substantial financial support? The answer is that they could not. By the time that we had finished all the various agreements, it was almost doomed to failure. The government subscribed money to UCS, but there are two things here. In the first place, it was not enough, and in the second place, it took so long to get it. We had the Fairfield Experiment, which did not help, because it raised the wages heavily. What really annoyed me was that Fairfield had taken on ships at ridiculous prices, which could not possibly be met, and we at UCS were left with them. The Fairfield Experiment was a disaster. In addition to taking on a container ship [*Jervis Bay*] at a much reduced price, Fairfield also agreed to build two survey ships for the US Navy. When I took them on at UCS, I found that the shop was full of bars being welded, whereas the bar could be drawn by miles at Colvilles steel works. All they had to do was go to the steelworks, and state their requirements. Meanwhile Fairfield's were paying very high wages, giving an overnight increase of fifty percent to Draughtsmen.

I thought the UCS work-in was stupid. There were various things that the men's leaders did like invading the Boardroom. Now; about this business about keeping people out of the yards, I went in and out of the shipyards and I was not stopped once. They published a statement [*the men*] that they were going to build ships themselves. Do you think any supplier would supply material for a gang of blokes who were staging a sit-in? Towards the end, the leaders became much more sensible. When we met with them one day, as I remember we told them that the men in the yard were trespassing, and that if there was an accident they would not get compensation. We spelt out the legal situation to them, and they began to see the sense of this.

On nationalisation, I do not think that there was any alternative in a situation of losses, losses, losses. No firm could carry on except Vickers with its nuclear submarines, Yarrow's and Vosper Thornycroft. Again, with Yarrow it was naval work, and Vosper got a lot of specialised naval work from abroad.

Interview No. 14: Anthony Hepper, Chairman, Upper Clyde Shipbuilders

I was a director of Thomas Tilling, and was on secondment as an Industrial Advisor to the Department of Economic Affairs in the Wilson Labour Government. During this period I was asked to join the Shipbuilding Industry Board, which had been formed as a result of the Geddes Report to restructure the British shipbuilding industry through grants and loans to encourage grouping of yards. There were three members, the chairman, William Swallow from Vauxhall Motors, Joe Gormley from the National Union of Mineworkers, and me. I was assigned to the Clyde by the SIB and in particular to see whether I could form the five shipyards on the Upper Clyde into one group. This was the formation of UCS out of Fairfield, John Brown, Connell, Stephen and Yarrow. When the company was formed the problem arose as to who would be chairman? There were no other candidates so I took the job on the basis that there would be a Managing Director with shipbuilding experience appointed. Unfortunately, we were unable to find such a person, but we did find Ken Douglas eventually. He came from Austin and Pickersgill on the river Wear, and was basically a shipbuilder, which I was not.

I was at UCS for the whole of its life from 1968 to 1971, a total of three and a half years. I was the only chairman of UCS, and my involvement in the shipbuilding industry ceased the day that I left UCS. As soon as I became Chairman, the SIB said that I had to leave them. At the same time Thomas Tilling, who were paying me, took me off their payroll.

On the SIB, the Geddes Report was the bible; after all, we were formed to implement Geddes recommendations. At the time I accepted it as being a well researched Report. In the final analysis I realised that it was not researched in the depth it should have been, and in fact that the problems with the shipbuilding industry were far deeper seated that the Geddes Report ever revealed. Had I known that, I do not think I would have become involved. Geddes did not get to the heart of the problem.

The industry was weak in a number of features. One was that the units were too small. Secondly, the management, in modern terms, was poor, and in particular, financial management was almost non-existent. Marketing management was completely non-existent. Thirdly, the labour forces were very old-fashioned, and conservative with a small "c." The restrictions and working practices made life virtually impossible. This was particularly so of gravity-feed welding machines, which other countries had operated by one man. We refused to have any at all. Capital expenditure was almost non-existent. We were totally outdated. All these things were not properly revealed in the Geddes Report. For example, there was not one single man in the five yards of the

Upper Clyde whose duty it was to go and talk to customers, and to gain orders. One yard; Stephen's had not taken an order for two years!

My first priority at UCS was to build a proper management structure, which we did. We employed marketing and financial directors. The second priority was to establish accounting procedures, financial controls, so that we would know what was happening. This gave rise to the problems of measurement, without which it was impossible to know what was going on. This had to be unified. We then had to address the problems of the workforce, where they had a very large number of different wage rates. There was something like four hundred different rates of pay, which had to be rationalised. We got them down to about seven, I think. We also had to try to get the trade unions to work together, which was a virtually impossible task, because they fell into two categories. There was on the one side, the Boilermakers, who formed roughly one-third of the labour force. They built the hulls of the ships, doing the welding and associated work. There were, on the other hand, the outfitting trades [thirteen *of the fourteen unions*] who did everything else, like installing the engines, electrical work, plumbing and so on.

The Boilermakers thought that they should have a shilling more than the others because they had the greater industrial muscle. The others did not agree. So you could never build a ship effectively because if you went along with one side, the other would go on strike. You would either have one-third on strike or two-thirds. This was a permanent feature. I set up a council to try and get them to work together, but the Boilermakers refused to sit down with the others, because they had refused to discuss wage rates.

As to the constituent yards; John Brown was part of the John Brown Group, which was a public company. They were completing *Intrepid,* an assault ship for the Royal Navy. More importantly, they were working on the *QE2*, which at the time had two more years left to completion. It was employing about 4000 men including sub-contractors. The yard had no further orders. Connell's was a family company, who built cargo liners mainly for the Ben Line. It might have been able to exist for a short time on its own because it had a very close working relationship with its customers. It was very old-fashioned. Fairfield was owned by various people including the Government and the trade unions. It was undergoing the Fairfield Experiment. They were trying to introduce what they believed were modern management methods to shipbuilding. But the increased overhead costs coupled with taking orders at low prices meant that the company was insolvent at the time of the merger. They did, however, have a great deal of work to complete. There were two survey ships for the US Navy, one guided missile destroyer for the Royal Navy, and a container ship [*Jervis Bay*] nearly as large as the QE2. In all, they had about two years work. Stephen's was a family company which was completing a Leander class frigate for the Royal Navy. When this was finished there were no more orders. There was no way that they could have survived.

Yarrow was part of a public company, but was dominated by the Yarrow family. They were short of orders, and almost wholly dependent on the Ministry of Defence. They strongly disliked the idea of merging with the others. That is why the idea was conceived of them being a fifty-one percent subsidiary company within UCS.

If it had not been for the loans and grants being made available in accordance with the Geddes Report, I personally would have recommended that John Brown complete the QE2 for their own account, and that the Fairfield Experiment should have been allowed to take its course. However, since we were implementing the Geddes Report, this was not possible, and would, of course, have meant that the other companies would have gone to the wall. I thought that we ought to have let those two run their course before grouping the yards.

There was a sense of complacency among the older shipbuilders, and a degree of arrogance too. I think they almost certainly resented us coming in from the outside, and I think the fact that I was not a Scot was not helpful. The dynasties were a major problem. If you form a company consisting of the five previous chairmen, with three of them becoming deputy chairmen, [*Iain Stewart, Charles Connell and Eric Yarrow*] who had previously barely spoken to each other, then it was almost impossible to weld them together into a team. They were fighting a tribal war with each other. It did continue for a bit on the new Board. There was a row with Iain Stewart, who then resigned. After that, it settled down. He was the main source of disagreement with the others. He and Charlie Connell and Eric Yarrow could not get on at all. This was partly because of the Fairfield Experiment and partly due to personal antipathy. They came from quite different backgrounds and they hated each other.

We then rationalised the workforce. The problem with building a ship is that there was little incentive to complete the job. The men would drag the thing out. As a result of that you had a financial problem as the progress payments did not come in on time. The crisis at UCS was almost continuous, and one of shortage of money. We were always kept short of money, deliberately, by the government, who did not recognise how much money was necessary, nor did we. This was largely because of the lack of financial knowledge, and the number of strikes and stoppages, of which there were a considerable number. Most of the time, someone or other was on strike.

The Labour Government, particularly Anthony Wedgwood Benn, kept us short of money, but less so for Harold Lever, who recognised the need for money. Benn regarded the Upper Clyde as a sort of social experiment. He used to come up with his own megaphone and address the workforce, which made it very difficult for management running the place, when you had the Secretary of State talking directly to the workforce. We were in constant need of working capital. It was also needed to finance losses. What money that was given was always too little and too late. This, of course, affected our relationships with our workforce, our customers and our suppliers. We were attempting to build up UCS as a viable unit. We were aware of economies of scale, and we were also going to reduce the operation. We were also going to produce a standard ship. We actually designed a Clyde ship to be built at Govan.

When the Conservative Government were elected in 1970, we were short of money, around £3m. John Davies, the new Secretary of State said that he was not prepared to produce that sort of money, and we said that we could not go without it. In the event, we produced about two and three quarter million pounds ourselves largely from owners who wanted to see their ships finished. Once we got this sum from the owners, we then said to the Government that we had done more or less as they asked us to do. They then procrastinated, and then, finally, pulled the plug. As soon as the Receiver was called in, I left. They, famously, made a u-turn some months later.

At its height UCS directly employed some 14,000 men. There must have been about twice that number employed indirectly, either as subcontractors or in another capacity. There was still far too many people employed and the productivity was pathetically low, partly due to low capital expenditure and restrictive practices. We would have liked to have two shifts working to maximise the facilities for as great a length of time as possible, and therefore recoup substantial overhead costs, but it was a social, and also a transport problem.

Interview No. 15: Harry Osborne, Civil Engineer

I was brought up next to the shipyards in Old Kilpatrick, and went to Clydebank High School, and was present at the launch of the *Queen Mary* and *Queen Elizabeth*. I went into civil engineering, and then went into a civil marine works, and at various times got involved with shipyard work. I worked on modernisation projects at Fairfield, John Brown, Scott Lithgow and Harland and Wolff at Belfast. However, it was mainly Scottish yards that I worked with.

There was some good management, for example, Ross Belch [*Lithgows*] and Derek Kimber [*Fairfield*]. However, there was a problem with the unions. At the time of UCS I was involved with some work at the John Brown yard at Clydebank. The impression I got was that in some ways management had lost control. There were more people walking around the yard than there were working, and the management were not doing anything about it. So the union problem was the main weakness, and had been present for a long time in the shipbuilding industry.

Although some companies such as Harland and Wolff and Scott Lithgow went in for super tanker construction, I think the British shipbuilding industry would have been better to concentrate on building more specialised vessels. The small-scale of individual firms was considered wrong, but small firms are not necessarily bad, and in any case, they can grow into bigger firms. Because they are small does not mean that they are inefficient. Lack of investment was possibly the case in many of the yards, but not in all of them.

On industrial relations I witnessed a lot of problems. At the back of Lithgows canteen, you would find a couple of hundred men being harangued by a shop steward. There was a certain amount of stirring up at times, which did not help matters generally. This was probably exemplified by the sit-in [*UCS work-in*] of the Jimmy Reid and Jimmy Airlie [*Communist trade unionists*] period. What was required was leadership and discipline. On UCS, what they did was essentially split the Clyde in two. To class John Brown and Fairfield together under the same management was wrong, because they were two entirely different yards that would have been better left on their own.

Nationalisation was unnecessary, and they would have survived without it. There is no doubt about it. Like all industries you have to be in the blast of competition, and must organise your affairs to cope with that blast. Nationalisation took the firms out of the real realms of competition. The industry would have found its own feet, it is like all businesses, you survive or you go under, it is a question of looking for opportunities in the world.

Interviews
The Tyne

Interview No. 16: G.H.R. Towers, Managing Director, John Redhead and Sons, South Shields

I was fortunate enough to have a chance of two family jobs. My father was an Average Adjuster, a shipping lawyer, and my mother was a Redhead, and I was asked if I would like to join Redheads. I served three years in the Drawing Office, and then went outside into the yard. The success of Redheads was that we had three departments. We built and engined the whole ship including making lifeboats. When shipbuilding was going down, ship repairing kept us going. We also did a lot of engineering work for other people. After seven years I was given the job of Assistant Manager. Then my uncle, Sir James Redhead died, by which time I had been made a Director of the company. The family decided that I should take on the job of Managing Director, but I do not now think I was ready for it. Redheads were a family company, which had a large proportion of people who had a father or grandfather working there. There was a Geordie saying that, "you canna get a job at Redheeds unless you see a coffin gan oot."

The mistake that was made, where I was concerned, was that I should have been sent to another shipyard for at least a year or two to gain experience. The strength of Redheads was that we built a good ship, and had a good reputation. I know for a fact that we got ships, although we were not always the lowest price. I became Managing Director in 1940, and held that position until Swan Hunter took us over post-Geddes [*1968*].

We specialised in cargo vessels, and had exceptionally good labour relations. There were about twenty-one trade unions in our engine works and shipyard, and I had difficulty when transferring men from one department to another as his wages had to be adapted to the department for which he was working. Generally speaking, we were a happy outfit, but there was nearly always too many people employed to do the job. We tried to install new machines, and in some cases we were not allowed to do so by the trade unions. The weaknesses were lack of space. We got to the stage when we were asked to build bigger ships, which we tried to do. We eventually got up to 30,000 ton bulk carriers, but the local Corporation never really helped us cash-wise, although when we bought the old Smiths Dock land, they filled it in for free.

Late delivery never worried me at all. To my knowledge we never had a glaring case of late delivery. Our productivity was average, but, at that time, I had nothing with which to compare it. We did try to break down the squad system, and in the 1960s employed a firm of consultants called Associated and Industrial Consultants. They were exceptionally good. They had a gentleman called Hodges working with them at that time, and when our Shipyard Manager died suddenly, because I was impressed with Hodges I offered him the job as Shipyard Manager. I gave him a contract for three years, and he was the man using a stopwatch. I did not realise he was using it in sight of the men. I received many complaints about this, and it became so bad that I went down to the yard. I was told by more that one person that they were being stopped from seeing me, and being watched by, "the Beast of Belsen." That same man wrote a very good paper for the North East Coast Institution of Engineers and Shipbuilders based on our attempt to break the squad system. However, it never really materialised at that time.

We bought new technology, but we could not afford what other yards could have afforded such as Vickers and Swan Hunter. We did, however, spend a lot of family money. We had some interesting ship owners who would not have welding on their ships when it came out, and we were asked to continue riveting when other people were welding. Poor management, I took the view that a happy yard was more important than having strikes or trouble. Redheads never had a strike, which started in the yard itself, in the history of the company. When I was President of the Shipbuilding Employers Federation, I used to travel to London with Ted Hill [*later Lord Hill*] who was the Boilermakers Society Leader. I asked him and his wife to launch a ship, and I was criticised for this at the SEF, but I did this deliberately. Ted Hill wrote a letter to me in which he said that if all the employers were like us, then the trade union leaders would not have a job.

We did have 2000 people working for us at one time, but it went down by stages to 800. There was, however, over-employment. Fitting a sidelight of a ship involved too many tradesmen: woodworkers, drillers, caulkers, riveters and plumbers. There was, therefore, a lot of unnecessary delay in performing a simple task. During the 1960s what changed was the manner of payment for ships. It used be all cash by instalments, and the final instalment was usually paid on the delivery of the ship. Then a loan system started, whereby we had to start borrowing money. Our relationship with clients, however, did not really change that much. I got orders almost entirely by myself. I took a team with me but never more that three people. I well remember going to Peninsular and Oriental with just the two people, myself and the Chief Draughtsman. We met the Japanese there, and their numbers were in the teens. However, we got the order, and the Japanese had a completely different outlook.

Labour relations in the industry generally were bad. Sometimes I would feel that the employers were bloody-minded, and I would also say the same about the unions. If I ever had any serious trouble in the shipyard, I would have been able to telephone Ted Hill and ask him to help sort it out. On management, my son, Rodney reckoned that our people were good, but when he went to Japan, he said that they were a different class. He also said that there were so many of them, and that the wages and conditions were completely different.

We had Board meetings every two months, and we also had the Works Committee. I used to handle the important matters and would call of the Shop Stewards in the canteen, and explain the situation to them. I always told them about any new orders, before the news hit the press, even if it was just a day before. We planned when we knew that we had to build a bigger yard. We had the Town Engineer visiting us very often, but we could never get the road closed, which would have made expansion much easier, and we, therefore, had to expand within by closing a berth, but generally speaking we had a good relationship with the Town Council.

I thought that there was no question that international competition was unfair. Nobody could really find out what was happening there. We do know that the British ship owners were getting by going abroad. They were receiving big money; perhaps the equivalent of the materials or the equivalent of the labour costs, but there was no hope of competing at all.

The thing that annoyed me about the nationalised British Shipbuilders was that there was not a practical shipbuilder on the Board, and when the employed Admiral Griffin as the first Chairman, I am told that the wrong Admiral received the invitation. The man who impressed me more than anyone else on the Geddes Committee was Joe Gormley [*Joe Gormley was not on the Geddes Committee but was a member of the subsequent Shipbuilding Industry Board*] he was a mining engineer and asked more practical and sensible questions than anyone else. Geddes, himself, did not impress me at all.

On grouping, I did go to London as Redheads wished to remain independent, and was told that if we did not, no government aid would be forthcoming. The proposals to group yards would have worked fine if people had done what they said they would do. When we were taken over by Swan Hunter [*the second largest shipbuilding firm in the UK and the largest on the North East coast*] it was fine for the first period, and then after a while I was told they wanted to dispense with the name Redhead, and then Swan Hunter became completely domineering, and insisted on calling us the South Shields Shipyard.

I do not think the later formation of British Shipbuilders was inevitable at that stage. The thing I did not like about British Shipbuilders was that there was no one you could talk to on a practical basis; they co-opted people on a part-time basis. The atmosphere in the industry changed very quickly, and people were changing at the top at the very same time. The attitude of people now amazes me. My attitude was to build a good ship, and to have a happy yard. We bought a welfare club for the men. During the interwar depression my uncle gave £5000 to the men at Christmas time out of his own pocket. That is the sort of family we were. I used to give money every year to the Bowling, Tennis and Cricket clubs. They do not want to know things like that these days. There was a very big change in shop stewards. Swan Hunter were jealous that we were such a happy company, and never had trouble, so certain people were being moved from Wallsend-trouble makers ended up with us. After grouping, orders became scarce.

Interview No. 17: J.E. Steele, Deputy Chairman, Swan Hunter, and Board Member of British Shipbuilders

In 1950, I was apprenticed as a Ship Draughtsman with Swan Hunter. On completion in 1955, I then spent a year in the shipyard as a journeyman; it involved three months in the Loft, two months in the Plumbers Shop and the Electrical Shop, and then time working with the shipwrights, carpenters, platers and fitters lining off. I then spent two years National Service in the Royal Air Force throughout the Far East, before returning to Swan Hunter in 1958 to work in the Steel Drawing Office. There, I calculated scantlings for steel and aluminium structures, and welding profiles ensuring that they complied with regulatory bodies' requirements. Around May 1959, I joined the Ship Estimating Office, which I took to like a duck to water. By 1961, I was appointed Section Leader, and two years later, Cost Controller, before in 1964 being placed in charge of all purchasing and material handling. In 1967, I was appointed Group Purchasing Co-ordinating Manager and Material Controller, and three years later, Purchasing Director. By 1974, I was appointed Deputy Chairman, Chief Executive in 1976, and Chairman and Chief Executive in 1977. Under British Shipbuilders I was a part-time Board Member, and in 1981 was appointed non executive Chairman of Cammell Laird, and overall Managing Director of the Composite Shipbuilding Division. In November 1981, I was appointed a full executive Board Member, and Executive Vice Chairman of British Shipbuilders Offshore Division. A year later I was appointed a Managing Director of British Shipbuilders [*one of four*]. In 1982, I had resigned from my chief executive position at Swan Hunter, but retained the Chairmanship in a non executive capacity. In May 1983, I was appointed non executive Chairman of Scott Lithgow. When Scott Lithgow was sold, and when Cammell Laird [the two large companies in the ill-fated Offshore Division] integrated into Vickers in 1984, I became Executive Chairman of Vosper Thornycroft, which was subsequently successfully privatised. Thereafter, in January 1987, I was appointed Managing Director of NESL for the last three months before its closure, and subsequently I was made redundant by British Shipbuilders.

Swan Hunter had a lot of orders in the late 1950s. However, the market had changed significantly. One of the dramatic changes during the late 1950s and early 1960s was the introduction of fixed price contracts, which gave far more importance to the accuracy of both prices and contracts, and, of course, the control of costs. This was more or less thrust upon the industry due to the Japanese. Whilst they had been recovering from the effects of the war, with considerable financial help from America, they were initially treated with suspicion, and were unsuccessful at building ships. In fact they had a number of examples of poor quality, and also had a bad name for meeting regulatory bodies' requirements. Therefore, to convince ship owners, they offered fixed price contracts and, I suspect, would more or less accept whatever specifications that were demanded. I believe that this was part of a national strategy. The clients, particularly with American influence, produced more and more stringent terms and conditions of contracts, until, eventually, the contract documents were actually thicker that the specifications with stringent penalty clauses for specified performances and delivery dates. However, the Japanese learnt fast and soon were leading the world with ship specifications and high quality standards.

In 1959, whilst I was in the Estimating Office, Swan Hunter started to grow, and began to centralise some of their departments. The Wallsend shipyard took over the financial and commercial management of the Neptune shipyard, which had previously been run as a separate subsidiary company. They then began to integrate various other aspects of the Neptune yard management into the Wallsend shipyard, followed by some integration of aspects of Barclay Curle on the Clyde.

One of the first integrations was finance and estimating, and this developed as I progressed to Section Leader in charge of all estimating by 1961. Contemporaneously, we were involved with two ultimately unsuccessful tenders to build a replacement Queen for Cunard in conjunction with Vickers Walker yard on the Tyne. The joint bid actually won the first contract in terms of design and price, but Cunard decided not to proceed with it, and the second bid was won by John Brown at Clydebank and resulted in the building there of the *QE2*.

As of consequence of various internal changes, we were finding that, whilst we were tendering at fixed prices, there had been no change in attitude within the yard, in terms of working to budgets, and of meeting cost parameters. People were still building the vessels as they had done previously, which was generally on a cost plus basis. They were buying their components from supplier's who were not terribly competitive, and the labour costs were not being monitored or controlled, and really the ships were turning out at whatever they cost. Because of this, Swan Hunter decided that they should appoint someone with the title of Cost Controller who would analyse the contract price of each ship, and issue certain budget target costs, and also ensure that those target costs were closely monitored and maintained. Very shortly after I was involved with this work, it became apparent that tremendous potential savings could be made in buying using the firm's overall purchasing power, distributing enquiries more widely, and insisting on fixed prices from suppliers. The Managing Director, Tom McIvor, in particular developed and supported this, and gave me target savings to achieve, which were achieved in the region of savings of twenty percent or more in total material costs.

The Swan Hunter Group comprised more than shipyards, it included ship repair, civil engineering, and also included a number of international operations. By 1967, I was appointed Group Purchasing Co-ordinating Manager and Material Controller with annual purchases of about £60m. Come 1968, we had a merger of quite a number of yards on the Tyne, John Redhead, which was in serious financial difficulties, and who had already approached Swan Hunter about a takeover, Hawthorn Leslie, which had financial problems due to their house building diversification, and Vickers Walker Naval yard. This resulted in the formation in 1968 of what was initially known as Associated Shipbuilders, and thereafter Swan Hunter and Tyne Shipbuilders [*SHATS*]. A local Board was created to run the shipbuilding business on the Tyne, and I was elected to it. Shortly afterwards in 1970, I was appointed to the company main Board as Purchasing Director, and became deeply involved in what was termed the Company Sales Taskforce led by Tom McIvor. We went all over the world, often at short notice, and were very successful at gaining orders.

Swan Hunter's major strengths were its management, its flexibility and its skills. Its weaknesses were labour relations and costs much higher than Japanese and European levels. I believe in actual fact that its progress was considerably inhibited by the number of experts and committees appointed by successive governments who were constantly being established to look into the industry, and the way in which it was working, and to produce reports and recommendations. We were also impeded by the commercial incentives that were offered by Geddes, [*grants and loans*] for example, to establish and construct [*by way of mergers*] very large combines, which meant that management that had been previously very effective in the more efficient companies, was stretched much further than practical to compensate for the number of lame ducks, which were saved by integration or mergers encouraged by the government. Swan Hunter was affected by these actions by stretching the talents of their managers over much wider fields, sometimes in geographical terms, but also because some firms which were integrated into Swan Hunter needed much more management attention. There was a tremendous amount of interference and constraints by either government or by government-sponsored or appointed bodies. There were a number of companies that were saved who should have, frankly, gone to the wall because of their performances in the market. However, market forces were not allowed to prevail. The only effect that this had, in my view, was to impede the effectiveness of the more successful shipbuilders, thus weakening British shipbuilding overall. In my view, Swan Hunter, and others, suffered from "artificial growth" and in accepting the incentives that were offered by government, encouraged over-centralising to help poor performance.

I think that there were a small number of facets within the Geddes Report which were relevant. Geddes certainly indicated one area which the industry had not concentrated on, and that was the effective use of its purchasing power, and the more competitive use of its suppliers. I think that Geddes was wrong in recommending that large is beautiful, partly for the reasons mentioned, *viz.*, artificial growth, stretching management's talents by over-centralising, but I must admit there is a considerable degree of hindsight here.

If we go back to when Geddes was written, it was not only shipbuilding, but industry generally, which was seen as having a solution to its problems in big is beautiful. I think where Geddes was wrong, not Geddes himself, but the interpretation and implementation by others who took a number of companies that were already in trouble and decided that their salvation was to bung them all together with more successful operators, and give out some goodies in the form of investment grants, loans or whatever, which, in turn, weakened the overall position of the most successful companies. I think government played a negative role in the industry, but I must not give the impression that everything in the industry was hunky dory, because it was not. There was a hell of a lot that needed doing. Some companies were further ahead in technology, industrial relations, capital expenditure and production techniques than others. Perhaps, because of that, I am very influenced with the view that in the period 1967 to 1971 that market forces should have been allowed to prevail in a natural way.

The industry was, in many instances, requiring considerable investment and cash, and if an industry which supports a tremendous infrastructure with a high employment factor requires support and investment, then, governments will get involved. It is in the nature of governments to want a consultants report before making any investment or giving support. They duly receive the report, and of course they will interpret it in order-I would presume-to suit their own political views. I think that is where it perhaps went wrong, although I think successive governments' intentions were admirable. I believe that where they went wrong was in their attempts to control expenditure they lacked the necessary commercial experience.

Look at the investment grant. It was meant to assist UK companies in the purchase of capital goods-like ships from UK manufacturers, but somebody did not notice the fact that the investment grant was still payable to British ship owners if they bought a ship in Japan. What therefore happened was that British ship owners, being commercially alert in my view, pressed the British shipbuilder by saying "ok, so that is your price, but we can get it for so much less in Japan, and the investment grant is common, so if you want the work, drop your price." Thus, the ship owner benefited from British taxpayers monies being paid as an investment grant for ships built in Japan, in addition to the loss absorbed by the shipbuilder because his price was too low. I believe that the total amount of investment grant sums paid in such circumstances amounted to approximately £500m. It would have been far more practical and commercially sound for the government to have allocated a fixed sum in grant to the shipbuilder for them to use in their discretion to what quotes they applied it. That is an example where the trend for the government to have more say in the industry was wrong.

Were labour relations and high costs connected? Partly; obviously you can not have bad labour relations and not have high costs. Labour relations are perhaps the area where in the 1960s and 1970s we did not progress as much as we should have done. We suffered, like so many other industries, like car manufacturing and coal mining, in that there were too many unions, shop stewards, and vested interests. Regrettably, I think that by the 1960s and 1970s the pendulum had swung too far in terms of the strength and power of the shop stewards, although there was, and always will be, a need for trade unionism to keep a balance and represent employees. Management in shipbuilding were in the difficult position in the 1960s and 1970s that if they stood up to the sweeping demands of the shop stewards and trade unions, resulting in strikes with their enormous under-recovery of overheads, then companies would quickly go bust.

At the very least, they would gain further bad publicity with unsatisfied clients due to late deliveries of ships caused by industrial action. I think that there was, however, a certain degree of abdication by management due to lack of support nationally. But, in saying that, I do not know who or what shipbuilding company would have had the funds available to cover the high costs of under-recovery of overheads involved in a long strike. After all, we found that, subsequently, in the 1980s, that whole issue appeared to me to be balanced by the government effectively funding the tremendous cost of re-establishing management's right to manage by establishing a more equitable balance of power in industrial relations by standing up to the unions more outrageous demands in the nationalised coal and steel industries resulting in long strikes etc.

Until nationalisation in 1977, labour relations in the shipbuilding industry were antiquated. It was very much a, "them and us" syndrome with deep-seated suspicions. Also, within the workforce themselves; were tremendous traditional differences and prejudices. You had the Boilermakers who had for generations been lords of all they surveyed, and looked upon other trades as being secondary. They believed that they, the Boilermakers, determined whether ships were built effectively and on time. They believed that they were irreplaceable, and they were, therefore, very powerful. They looked upon themselves as being the top skilled men, and that therefore they should have the top wages. They were very traditional and resistant to change. They overlooked that there had been tremendous developments in the design and construction of ships, and that we now had far more sophistication in electronics, hydraulics and mechanics with far more emphasis on the work of the outfitting trades. Therefore, not unreasonably, the outfit workers asked why they should not have the same basic wage rates as the Boilermakers. They had served their time just the same as had the Boilermakers. We had the most ridiculous situation in the early 1970s where, with high rates of inflation [*at one stage, twenty-five percent!*] you had wage rate leapfrogging going on between steel, outfit and ancillary workers, each negotiating separately. Thus, you had wage rates being increased four or five times per year. The unions would not agree to a single date for negotiating collectively, and insisted on negotiating separately. Historically, the ancillary workers [*labourers and semi-skilled operatives*] were looked upon as the guys who did what they were told by skilled people. However, this was changing, and they were also proving how indispensable they were. The Tyne was particularly bad at these wage differentials, which were not resolved until 1979, after nationalisation and two strikes which lost the Polish ships to Govan Shipbuilders on the Upper Clyde.

One positive contribution that did come out of nationalisation, in my view, was uniformity in terms and conditions of employment and the harmonisation of labour relations generally, but most of all, uniformity in particular of basic wage rates for all staff, skilled and unskilled personnel, nationally were concerned.

There is no doubt about it that in shipbuilding, the Japanese, man for man, were more productive that the European, and in turn, the European worker was more productive that the British. If the Japanese were earning the same amount per day as our skilled men, then they produced more in that day than we did, and with fewer men. Thus, their overall costs were much lower. That level of productivity was not only relevant in the cost of building the ship itself, but also in the manufacture of all the components which were required to complete the ship. So the overall total effect was a significant reduction in cost of construction, and thus price, that we in Britain could not compete with. It was not just purely and simply, as people had said in the past, the fact that the direct labour costs were lower, it was also the cost of the raw materials, equipment and consumables used in construction of ships which were lower. It all reflected right the way through to the prices for services provided by subcontractors, and goods from suppliers and so on. It was not only British shipbuilders who were uncompetitive, but virtually all the marine equipment suppliers when compared to the Japanese.

Swan Hunter's marketing policy was based on the fact that it was a mixed shipyard building naval and merchant ships. We recognised if we were to utilise our strengths it would be folly to try and build standard merchant ships. We would never be able to compete because of the higher initial overheads and standards of quality control documentation we would have to carry as a builder of naval vessels. Naval vessels demand a far higher level of skill content, and tend to be more costly than standard merchant ships. We did also recognise, however, that with naval vessels you have a much longer term building programme on which you should be able to spread and recover your technical overheads, particularly on the parallel building of more sophisticated merchant ships with shorter building periods. This proved to be of benefit to the Ministry of Defence and to Swan Hunter, who, as a consequence, could quote more competitive prices because of the spread in overheads. Because of this, Swan Hunter did build a very wide range of sophisticated merchant ships that did require all the staff and skills required for naval vessels.

On international competition, some of it was unfair. I think that it seemed mainly unfair because we were in the same international market place, but felt that we did not enjoy the same degree of support. We believed that the French had a unique method which took some years to puzzle out. We never ever knew how they were able to offer some of the low prices they could until we suspected that they had a unique system within COFACE, which was similar to the UK Export Credit Guarantee Department. COFACE was effectively a credit insurance organisation, but in addition we believed also offered cover for inflation. COFACE was initiated and supported by the French Government. It was never directly established that cover for inflation was available, but it seemed the only explanation as to how French builders could quote fixed prices for ships at apparently current day levels. If all inflation was recovered from a source such as an insurance company for a very small premium, this could have made a very great difference, if such an arrangement had been made available to us especially in the mid 1970s when we were facing inflation at levels of twenty-five percent. When we first used to quote fixed prices in the 1960s, inflation ran at around two to 2.5 percent per annum. It also became apparent that some of the credit and financing deals being offered by other countries were imaginative to say the least. Indeed, some of the interest rates and tenure of credit terms were such as to appear to make the price of the ship inconsequential.

At first we thought that we could limit Japan, but it was not to be. What Japan did was just simply to build massive new building docks which could accommodate ships of up to 750,000 tons. There was no way that we in Britain could fund such capital expenditure, and we can only assume that indeed the debts that Japan had accrued in the first tranche of construction were written off, and, or, the funding amortised. However, a significant contribution to the survival of merchant shipbuilding in Britain would have been if a satisfactory formula could have been found for both the ship owner and the shipbuilder, which would have enabled the British ship owner to place between twenty and twenty-five percent of their annual new tonnage in British shipyards. This should not have affected their cross-trading, and it certainly did not affect the French or Japanese who placed one hundred percent of their needs in their own shipyards.

Late delivery was quite a major problem at Swan Hunter. We did our best to justify it in terms of *force majeure* [*strikes, inclement weather conditions, etc.*] but we did have problems. Some of the problems in terms of quality and delivery were much worse with the suppliers than with the shipyards themselves. However, in the eyes of our clients as the problems manifested themselves in terms of the delay in the contract delivery of the ship, which was the responsibility of the shipbuilder.

Quite a lot of British shipyards, to my knowledge invested their own money very heavily in modern shipbuilding steel panel lines way ahead of time; for example Scott Lithgow, Swan Hunter and Cammell Laird. There was also very heavy investment in cranes and other equipment. I also believe that there was also some unwise investment at Hebburn, and on the Wear, but that view comes partly from hindsight. At Hebburn, Swan Hunter bought the old Palmer's ship repair yard in 1973, and built it into a shipbuilding factory where we planned to produce 100,000 tons per annum of finished steel throughput to produce two and a half 100,000 ton tankers or bulk carriers each year in that big dock. Our philosophy in buying the site was that it already included the biggest merchant dry dock on the North East coast, requiring us to buy only heavier lifting cranes, and erecting a steel shop or factory for which there were grants and capital allowances available. The market in 1971 indicated a demand for large tankers and bulk carriers which did not seem as if it was going to end. Therefore, it seemed a good investment to build a shipbuilding factory at Hebburn. It took three years to develop, but no sooner did it come on stream than the market for large tankers and bulk carriers collapsed. So it proved to be a very unwise investment, which never produced a fraction of what it should have done.

They were also very heavy investments on the Wear, particularly on covered-in construction halls. I must say that personally it always seemed to me that these investments were in the long-term unwise as they seemed to be upriver of the bridges [at Sunderland] and the river curves. They were on the wrong side for growth. Natural growth would tend to indicate that if you put a roof on something that it is going to be an impediment and a size constraint at some time in the future. Furthermore, everything organically, and ships in particular, grow in size over the years. Both Pallion and Southwick [*the two sites of construction halls*] had serious limitations in terms of draft [*dredged channel*], height above the water line [*bridges*] and length [*curve of the river*] and were limited to about 35,000 metric tonnes deadweight.

When I first joined Swan Hunter they had three covered in berths where the old *Mauretania* was built in 1906. However, once ships had reached

50,000 or 60,000 tons deadweight, five ton berth cranes were of little use and the constraints of height too great, so we knocked the whole thing down and built two berths that could accommodate either two 100,000 deadweight ton ships or one 250,000 deadweight ton tanker diagonally [*with the aid of the Shipbuilding Industry Board*] far less constraining and far more flexible than covered-in berths with their constraints in size and their ventilation problems.

In my view shipbuilding had a pretty awful lobby, and was very ineffective in comparison to the shipping lobby. I know that perhaps some of my peers would disagree, but we had a poor image particularly in the eyes of government, and could not stand up to the tremendous strength of the shipping lobby. Even now the British shipping lobby is tremendously strong, and very influential, but if you got back over the years then the UK shipbuilding industry, in my view, were pawns to be used in the overall commercial shipping lobby by the General Council of British Shipping, and could, and were, treated as hostages to fortune. Some British shipowners were absolutely superb and stood by British shipbuilding through thick and thin, and sometimes with tremendous embarrassment to themselves. Let us just say that others were not so accommodating.

I think some of the poorest support for shipbuilding, and the most divisive, was in fact the application of the Intervention Fund. It did not even make tremendous sense. Shipbuilders had to demonstrate definitely that before intervention was granted what they were quoting for the ship was absolutely break-even, a very difficult, if not impossible task, and if they actually made a profit on the ship they would have to pay it back. This meant that at the same time as one was trying to quote the most competitive price to win the work against the most cut-throat competition you had so many constraints and disciplines pulling you in so many directions at once, that you really could not function.

Going back to 1974, [*OPEC oil price hike, which first tripled and then quadrupled the price of oil*] the industry was facing one of its most serious problems with ships being cancelled left, right and centre. During my time with the shipbuilding industry there had been four occasions when orders were just not available, 1958 to 1960, 1967-1969, 1974 to 1978, and 1985 to 1989. In 1974 we had the Middle East crisis and we had cancellations at Swan Hunter of thirteen ships, which was traumatic. At the same time inflation was rampant, and capacity was unbalanced, because, as a consequence of Geddes, we had absorbed companies and yards that were less efficient, and which stretched our limited management resources. We were in a parlous position with over-capacity and over-manning, which we dealt with by redundancy. We also faced the possibility of nationalisation.

I recall that there were two things which helped Swan Hunter to get through this quite turbulent period in terms of management. Firstly, we successfully concluded an order with the Iranian Navy for a contract worth over

£40m and a build period of four to five years. It was for a new generation of large ships, ahead of its time. It was breaking into a market and a country, which at that time was everybody's darling, and had all the money in the world. Furthermore, it was a cash contract. The other important event was the successful conclusion of an enabling contract with the Ministry of Defence for a through-deck cruiser. An enabling contract was a reimbursable contract which enabled us to do certain technical work, and to compile a fixed price estimate for the construction of HMS *Illustrious,* [*aircraft carrier*] with an option that, if after a year we performed satisfactorily, and our price was acceptable, then we would be able to sign the construction contract at an agreed price. Apart from these two events, we would have been in a parlous position, as was the rest of the industry.

However, in addition to closing Redheads, and also reducing our workforce at Swan Hunter, we also had to settle claims from suppliers re the thirteen cancelled ships. 1974-1977 was a busy period. Furthermore, we were able to sell some of the cancelled ships we had finished, and three bulk carriers we had designed using surplus material and equipment from cancelled ships. We also had super tankers to complete and sell, which we achieved in 1977. HMS *Illustrious*, followed by HMS *Ark Royal* [*aircraft carrier*] and the Type-42 Destroyers with the three bulk carriers saved us. We already had the Type-42 contracts before the bottom dropped out of the market in 1974, but the two aircraft carrier contracts saved Swan Hunter as it is today.

Nationalisation not only seemed inevitable, in 1974 it seemed preferable. But in Parliament we went through this hybridity pantomime, if ever there was a case of the powers that be fiddling while Rome burns, then that was it. That absolutely tore the guts out of the industry because, as stated, we had a disappearing market with ships being cancelled with abandon, particularly tankers. We had costs escalating through rampant inflation. We had overcapacity in relation to demand and the need to reduce the number of people employed, with all the costs and difficulties with labour relations, which were entailed with redundancies. There was an absolute need in those circumstances to get more work and quickly. However, we, along with the whole world did not know whether we would be nationalised or not. We had ship owners, who I do not criticise, who were understandably reluctant to deal with us and place orders as they did not know if they had any guarantees, or who would own us. Understandably [*because of the likelihood of nationalization*] our parent companies would not invest in us. After all, you do not paint your house if it is going to be compulsory purchased.

We experienced a period of two and a half years where nothing happened [*on nationalisation*]. We stagnated and died while the unions were considering who would manage and who would not. Shipbuilding was haemorrhaging, and in my view was the political shuttlecock between the two parties [*the Labour government and the Conservative opposition who were ideologi-*

cally opposed to nationalization and fought the proposals tooth and nail] without consideration of the results. In my view, if someone tried with the most fiendish mind to find some way to destroy an industry, or strike it a fatal blow and say it was nothing to do with them; they could not have done any better. I personally think that the industry never recovered from that experience.

One aspect of impending nationalisation, which I welcomed, was that the rough capital costs of ships, depending on size and type was, say around £15 million in 1972-1973. Swan Hunter were producing about twelve of those per year, and we had clients who were saying that if they were going to place a contract with us for such a ship, then they wanted either parent company or bank guarantees, because we required eight to ten percent, £1.5 million, payable with the order, and another payment when we started to order steel, which were the normal terms of payment in our contracts. However, our parent company was only capitalised to about £14.5 million, which was only equivalent to one ship, and we required at least four or five per year. Such guarantees were not possible.

What could we do-go to the market with a rights issue? Who was going to buy any shares in shipbuilding? Furthermore, a bank guarantee is a liability which would reduce the asset value of the company in the accounts. Do not forget that the industry had constantly had a very high public profile, with criticism and a poor image given of its management and industrial relations, not the sort of image projected to encourage investors. I digress, but there is nothing more galling than to be sitting in some potential client's office in, say, Lagos or somewhere else abroad waiting to go in and hopefully clinch a deal and see a copy of *Lloyd's List* with headlines saying, "Another strike at Swan Hunter." It is of no help that the strike could have been on the ship repair side of the business, because the client does not know this.

To return to the issue of the limited capitalisation value of the company, and the requests for bank guarantees, nationalisation was the answer because no private UK shipbuilding company could provide. If someone was going to place a ship contract for £15 millon and were paying instalments in advance-shipbuilding has always been cash positive- it was understandable that they wanted a guarantee that the company would not go bust, which, of course, nationalisation and government ownership offered. If nationalisation had taken place in 1975, in my view, some of the yards which have closed latterly, would have been saved.

When nationalisation arrived in 1977, three years had been lost [*since it was first mooted by the Labour Government*] with little or no new orders being placed in the UK. So, whereas in 1974-1975 you only had the weaker UK yards in dire trouble, by 1977 everybody was in trouble, and UK shipyard management were having to do the things that should have been done three years earlier-like reducing the numbers of people employed and closing some of the older yards, whilst trying to win desperately needed orders to provide some work for the under-utilised shipyards and engine builders.

The other problem with nationalisation was, in my view, the government and respective Chairmen of British Shipbuilders never got it clear whether they were functioning as a centralised or de-centralised operation. Why they did not grasp the nettle and function as a centralised operation I never knew. As far as government was concerned there was only one shareholder, the government, and one Chairman, one Chief Executive and one Board, and thus everyone had only to answer to them. Therefore, there was every practical and economic sense to function as a centralised operation. Instead British Shipbuilders committed the cardinal sin of indecision and incurred the costs of being both centralised and de-centralised with multiplicities of Boards, managements, Minister's without Portfolios etc, and all the inefficiencies that go with it.

Because of these circumstances there were introduced into Head Office an awful lot of people who knew little or nothing about ships or shipbuilding. Indeed, many were "rolling stones." Not that lack of shipbuilding knowledge in itself is bad; indeed, the industry did suffer from far too much incestuousness, and there was a need for the introduction of outsiders, but not to the extent that happened. It seemed for a time that anyone that who had a knowledge and experience from within the industry was regarded automatically as inferior to someone who came in from the outside. The lack of strong organisational structures led, in my view, to the proliferation of staff and departments, sometimes only to answer the many and varied queries and exercises from government, and the multiplicity of Boards and management leading to diluted management control and communication with the resultant high costs.

In my view, nationalisation never really affected the warship building firms due to the lack of central control. They never did become a part of British Shipbuilders other than in name. They functioned totally separately as trading companies and seemed to be encouraged to do so by the Ministry of Defence for understandable reasons of competition. They could, and should have made a greater positive contribution to British Shipbuilders. I accepted nationalisation as essential and with an open mind. However, I became disenchanted and disillusioned as I realised that government are worse than anyone else in managing a commercial operation.

Looking back over three decades, I would say that the natural effects of market forces should have been allowed to prevail to the maximum extent during the early to mid 1970s. Had that been the case, I believe we would have now in Britain a bigger, stronger and more virile shipbuilding industry in better shape to compete and survive in the world today.

Interview No. 18: Peter Milne, Managing Director, Swan Hunter, and Board Member of British Shipbuilders

I left school at sixteen and gained a student apprenticeship in an engine building firm within the Swan Hunter group of companies. I ended up in the Technical Department, went to sea for a while, and came back to the Technical Department at Wallsend. Post the grouping on the Tyne under the aegis of Swan Hunter [1968] I was eventually moved into general management and naval architecture, and from there progressed to deputy managing director and then, three years before nationalisation, to managing director. In 1977, I was seconded to British Shipbuilders Headquarters, becoming a Board Member in 1981, first for engineering and later for the composite merchant shipbuilding division. Later on I was Board Member for ship and engine building, which covered all that remained in operational terms after the warship yards had been privatised. Ultimately I left in 1990, having completed almost thirty-nine years in shipbuilding.

The strength of Swan Hunter was volume of output and technical capability. We ran six shipyards, and we built a wide variety of ships, and remained profitable up to nationalisation. In common with a lot of other shipbuilders we did not realise that we were going to have to get smaller. Shipbuilding has tended to follow low labour costs, because it is labour intensive, and the cheaper you can build the more work you will get. We did not perceive quickly enough that foreign competition was going to have an impact on us. It never needed to be as bad as it became. It became a political problem. The survival of shipbuilding capacity became a question of political will. So we had countries like Italy, Spain, and West Germany who believed that shipbuilding was an important part of their maritime infrastructure, and who have kept fairly substantial shipbuilding capacity. Unfortunately, we took a contrary view, and we have not got a great deal left.

The weaknesses that we had are that generally we did not have enough talented people in management. Rather belatedly, firms started to train and recruit managers who were graduates, and who were well qualified, but the talent was thinly spread. The other area that we were weak was at the supervisory level. People had been promoted from being very good tradesmen, but were not very good foremen. It was not their fault, but managements' fault. It had not been seen quickly enough that we needed to provide training, and that we needed a different kind of supervisor. We needed to bring people through who had some sort of technician qualification. The people we had were not able to step back and take a view of planning, of quality, of organising, and all the things that are fundamental to being a good supervisor. To be fair to them the training was not provided until later. One of the things that British Shipbuilders did on a grand scale was to train people. The industry had always been extremely good at craft training, but we had not trained for flexibility and interchangeability. We tended to continue to produce the people we complained about.

We were also weak in marketing. We did not perceive quickly enough that the market had completely changed. At one time when you needed a new order all you had to do was get on a train to London, or go to one of the main maritime centres, see the people you knew, talk to them and begin to negotiate for a contract. The marketing was done by the chairmen and managing directors and that was quite adequate because they could get all the work they needed. However, when the market changed, we had to go round the world and persuade new people in Hong Kong and other centres. We were not equipped for that as we were dealing with unknown people. We had no marketing kit, no brochures or anything like that. We had a reputation which we traded on for a while.

Before the business was very much a customer-client relationship but that changed in the late 1960s to early 1970s. Price became an important parameter. The Japanese began to assert themselves in the market and we did not respond to that, feeling, perhaps, that it was something of a passing phase. There was not the complacency that people in the business were often accused of. They were working desperately hard, and a lot of ships were still being produced. In a way, one got a false sense of security. The work was continuing to come in and rather belatedly people saw that they were operating in a different market. In the international market we were faced with sophisticated competition from the Japanese. It was not an accident. Japan actually had a national policy to target particular industries and shipbuilding was one of them. It was a way for the Japanese to build up heavy industry. There is such a diversity of skills needed in shipbuilding that it is a good basic industry to start with. They worked at it, invested money, researched and developed, subsidised in direct and indirect ways to get them into the business. They were also very competitive. In those days the Japanese had very low labour costs. There was no negligence on our bosses' part. The market changed progressively, and we began to run out of work. Another problem was that our prices were too high. This was true of a lot of western European countries; prices were too high and orders dried up. As Japanese labour costs were so low at the beginning, they were inevitably going to make a significant market penetration. Later on there was unfair competition. When we reached the 1980s we were looking at heavily subsidised prices from Japan and South Korea. There were significant cutbacks in Japan, which showed that they were suffering, but the cutbacks were not as big as was the case in Western Europe. Two of the big Korean groups went bankrupt with enormous debts, and this was all because prices were too low. So yes, later on, there was unfair competition, whether it was deliberate or not I do not know. I have always had a view that the Japanese are looking for world dominance in shipbuilding, and are looking for at least a fifty percent share of the overall market. They could have been buying market shares. They may have been prepared to lose money in the short-term to eliminate competition.

What we in Britain should have done was to have taken a view as to the size of the industry we could sustain. We should have merged it into bigger groups so that we spread the overheads. As we approached nationalisation we had some very weak companies in terms of management, performance and accumulated debt. If the decision to let them run out had been taken then we would have lost the inefficient capacity, and we would have concentrated on two or three bigger groups. In reality that is all we had the management for. We should have concentrated on the technical lead, which at that time we still had.

By 1977 we were nationalised and the name of the game, unfortunately, was employment. There were proposals put forward by British Shipbuilders to rationalise more radically in the early stages. They were rejected on the grounds that part of our job was to keep people in employment. That was the political imperative at the time, and led to the Polish deal and nonsensical arrangements like that which were commercially disastrous. By the time it becomes a disaster the politicians have moved on to something else, and are not the people then taking responsibility.

On the point of British Shipbuilders being too bureaucratic and having inexperienced shipbuilding management, you have to be careful who says it. Some people were always against it, and never accepted nationalisation. Another thing that people did not understand was that we were initially running an organisation with 87,000 people in it. We were also trying to introduce a lot of new ideas and techniques, and training. Some people did not think all that was necessary and just wanted the money to come in, and to be left alone. I remember the early battles with these people who did not want to send a financial package to Reg Arnell, who was the financial controller in those days. Reg wanted monthly financial statements, profit and loss statements, rolling cash forecast etc, but some companies within British Shipbuilders had never done anything like that before. They just wanted to wait for the auditors to come in at the end of the year to tell them what had been happening. On the operational side it was the same for many companies. So a lot of things we were doing were imposing what were basic business disciplines on people who objected strongly.

What they were looking for was a very loose federation of companies, where we [*British Shipbuilders*] effectively gave them money. It is interesting that the people who were cultivated most in headquarters were in two categories. The first were the people in finance who allocated subsidies, and also the people in finance who dealt with capital expenditure. Anyone else who wanted a discipline, to look at improving productivity, or getting proper financial returns was extremely unpopular. But, I did not think it was unduly bureaucratic. The other problem was the way individual companies were run. What happened in shipbuilding was that the focal point of the firm was the Managing Director, and a lot of them believed that everything had to go across their desks, or alternatively, everything that went to British Shipbuilders had to go across their desks. They wanted to screen and filter the information, so they invested a lot of their time in doing that and consequently blamed the bureaucracy. Had they delegated, things would have run a lot more smoothly. There were, however, people who were brought into headquarters whose motives and objectives were deeply distrusted. There were people who spent a lot of money and did not appear to have much control. But I do not think it was worse than in any other organisation.

On privatisation, we would have been smaller in any case. Again you come back to the political aspect. The view taken by the board members was that if the price of subsidy and continued operation was privatisation, then so be it. Graham Day came in with a very clear brief and was very straightforward with us, telling us that he had been sent in by the Conservative Government to privatise the warship building firms. He wrote a Corporate Plan based around that. It was a mistake in that he should have allowed firms to be mixed naval and merchant builders. There should have been fewer groups, with a number of companies in the group with a headquarters each. Those groups could have been building a variety of ships, including warships in the same group. I agreed in the separation, but think that it was achieved in the wrong way, by privatising and distancing the companies from the rest of the industry. Some of it could have been privatised; it should have been done selectively.

Another thing I strongly believe that the politicians got wrong was how the industry was sold off. I do not think it should have happened a bit at a time. We got to the stage where the industry was too small. That is one of the big threats now. The industry is too small. There is not the critical mass whereby you can, for example, share and fund research and development. I believe that more effort should have been made to sell off blocks of the industry as an integrated operation. That would have left them with a better prospect. The strategic overview was lacking, but we were handicapped, certainly in the latter years. The Government did not want to hear our advice. For them it had to be privatised, and it did not matter how.

Interview No. 19: P.D. Christie, Managing Director, Swan Hunter

During the interwar period my father was Chairman of Swan Hunter and Wigham Richardson, so with this family background I had never wanted to do anything else but to become a shipbuilder. I went to Uppingham School, and when I finished there at the end of 1930, I went to Hawthorn Leslie as a pupil apprentice. I was moved across to the design room at the Neptune yard, and then became an improver for one and half years and spent that in the Estimating Office. During this time I went to Durham University to take a degree in naval architecture. I then went outside for experience as an assistant shipyard manager, before going on to take charge of the Design Office. During the war I became a Director of Swan Hunter and eventually ran the Neptune yard. In the late 1960s I was asked to inaugurate a training company to cover young apprentices, and eventually, adult training, foremen and management. I stayed there until I retired, just as British Shipbuilders took over. I was a founder member of the Shipbuilding Industry Training Board.

The great strength, and, in certain circumstances, the weakness of the industry in general was that it was made up of comparatively small firms. They were all run by individualists who knew quite well that the way that they built ships was the right and only way, and in many cases that was a great strength. The weakness there was that, excluding the war years, the shipbuilding industry had three good years and seven bad years. This meant that during the three good years you had to earn enough money to keep yourself going over the seven bad years. Some of the smaller units found that very difficult. This meant that you did not have all that money to spend on new equipment. You will find that where they did there were a number of firms who went to the wall because they had equipped their yard too well despite the fact that they could get work. When they came to settle their periods they were paying so much on the equipment that they eventually went to the wall.

The great strength otherwise, was that they were family concerns. By family, I mean that the whole yard was a family. The Managing Director knew everybody and they all knew him, and people had a loyalty to the firm. This was the case up to the 1960s. One of the things that killed it was the legislation about hiring and firing. It tended to take that family feeling away. For example, the foremen were the backbone of the yard. If you wanted anything done you called the nearest foreman, and he went and got hold of the man and the job was done. The foreman had the right of hiring and firing, and had a large say in the enrolment of apprentices. It was the most important job in the yard. When the legislation about hiring and firing came in the foreman's responsibilities were eroded, and they were never the same again. You then had assistant managers for everything.

I think that management tended to fall to pieces to a certain extent with the arrival of all these assistant managers. Some foremen were effectively put out of business. In addition, some of the new management were not up to scratch. When you had a managing director and a first rate shipyard manager, and perhaps one assistant manager, management was not poor.

I think that the strength in our firm was that the Neptune yard built the small ships and Wallsend built the bigger ships. That mean that we could have the machinery for the type of ship we intended to build in each yard. It meant we could cover the whole gamut in size, and be reasonably priced at the same time. I doubt that there was obsolete technology. There was in the smaller firms. In Swan Hunter we ran the various parts separately. To a large extent, the director of each firm ran it the way he thought it ought to be run. As long as it satisfied the Board that it was doing what it should do, then that is what happened. I would say that we were reasonably up to date given that we could not do the best thing, which would have been to build a new yard. If you wanted to make a very big change it interfered with building ships, so you had to do it in the right order. That applied especially to the period after the war when we did rebuild our yards, but we had to do it when production was on stream. I think that is where the Japanese and German firms who had to start from scratch had an advantage.

One of the things was that in Japan, the shipbuilding industry was looked at as being a very important industry, showing the world that Japan was recovering from the war. Also, Japan had a whole lot of people who were scratching a living and if they got the chance of getting into a shipyard and receiving the appropriate training, they were very happy. Low wages to them were riches. That was what really put Japan on the map. Later on when people began to find their feet and demand more money, Japan found itself fighting Korean competition. For twenty years they had the advantage, though. The Japanese could lay out their yards in the best possible way to give them the easiest way of working. The Japanese looked on shipbuilding as a prestigious industry, which our country did not.

I do not think that the Japanese government subsidised their industry in an unfair way, but they did in France. We were trying to buy [*tender for*] a passenger ship from France. In the end we found that the prospective French owners had been told that whatever anyone else offers by way of tenders they would be undercut, and that the French government would see that they were no worse off for it. In Germany, we found that if there was unemployment in Hamburg, the German government gave Hamburg a certain amount of money for the alleviation of unemployment. If the local government [*Lander*] wanted to give it to a shipyard so that it could undercut other people and get orders then that was up to them. When the German government was asked if they were subsidising shipbuilding they denied it.

Later on, all the talk about nationalisation did not do the industry any good at all. A lot of top people were looking over their shoulders wondering what else was for them. The drain from the shipbuilding industry began then. What very nearly wrecked the British shipbuilding industry was the nationalisation bill. The Conservatives [*the opposition party in Parliament*] were just as much to blame as anyone else. The bill was stopped as the various political parties scored points against each other. Anyone who could get out of British shipbuilding in that period did so. It was terribly damaging. By the time we got to nationalisation there was no going back. It was a mistake in the beginning. There could have been other ways. I think a certain amount of rationalisation in the industry, and examination of the money that was available, perhaps some help in re-equipping. On nationalisation, the industry was left with a top heavy management organisation which the shipyards had to pay for. It made it more difficult to get orders when it came down to price. The way it was done was a disaster. You can not do all your designing in one place when you do not know how various parts of the workforce thinks in various parts of the country. They could have had a small highly developed elite control, but not have a huge headquarters like they had. It was not nationalisation that caused it, but the interminable delay in the bill, and the fact that it was not thought out properly.

Interview No. 20: James Jardine, Finance Director, Smith's Dock, North Shields, and Finance Director Swan Hunter Group

I was invited to join Smith's Dock as Secretary in 1956; and in 1960, I became the company's first Finance Director. We had two yards, one entirely for repair work, and the other also for repair and new building work. We also owned the Grangemouth Dockyard on the East coast of Scotland. When Smith's merged with Swan Hunter, [1966] and later became the Swan Hunter Group of shipyards, I became Finance Director of the entire concern and remained there up to nationalisation

Smith's up to the mid-1960s were a happy company. They still had Sir Eustace Smith as the chairman, and although his family were not by then big shareholders, they had the best qualities of a happily managed family firm. They also had the advantages of a public company. We made very good profits at that time, two to three million pounds in most years. This was the period 1956 to 1964. When we merged with Swan Hunter, the strengths were that we were building, on average, twenty-four ships a year of different types. So we had access to the market for all types of ships. There was not anything, other than submarines, that we did not build. We got involved at Smith's in building oil rigs, but lost money on it. At Swan Hunter we consistently made profits when we came into the very big tankers. We had a lot of trouble with the Government, and the only people who built ships of that size, 250,000 deadweight tons, were the Japanese. We had an awful argument with Tony Benn. He said that if we did not build one of these ships then he would cut off the finance which was available through the 1972 Finance Act. This was in 1974, when he was Secretary of State for Industry. By that time, we had constructed a berth to build a 250,000 ton ship. So we were capable of building the ships. In fact, we did build quite a few of that size. The first one was in 1969 and was named, *Esso Northumbria*.

This was the argument with Government. We had followed the Japanese with regard to prices, but the Government were not prepared to finance the ship owner beyond the prices being paid by the Japanese. The Japanese lost a lot of money during that period. The first ships were being sold at around £5m and were costing something over £10m. We got caught up in that. If we had not been so involved with the government under the financing arrangements, we might well have just walked away from it rather than take these losses. To be fair, we did not know how things were going to develop. But when we got into building super tankers that was a whole new ball game. We needed all sorts of sophisticated machinery, and we lost money on the building of these vessels.

The sad thing about the 1972 Finance Act was that owners could go anywhere in the world to have their ships built and enjoy this subsidy at a guaranteed rate of interest. They would not give any subsidies to the shipyards at all. [*In fact, the government did introduce a system of tapering construction grants for a three-year period. Shipbuilders also received assistance in the form of rebates on indirect taxes and exemptions of supplies from import duties. Shipyards in development areas, as most of them were, were eligible for regional development grants, regional employment premiums and refunds of Selective Employment Tax*] I do not think that British ship owners let the shipbuilding industry down, if anything it was *vice versa*. Clearly the management of yards could have done more, so could the unions, the sub-contractors, the suppliers etc. We found it easier to buy a Kawasaki main engine for a super tanker rather that to get one built by GEC or whoever. I can think of one turbine driven super tanker where the main engine was built by GEC, and it was months late. You can not complete a ship without the main engine.

I think that the industry tended to be slated. It had been such a big industry in this country in relation to the total manufacturing output. It was slipping down the scale. There were new people coming in. At one stage, Sweden was being talked up as a very modern country for shipbuilding, and so it was, but they are nearly all bust now, and they spent untold millions of Krona. I have one or two friends in Sweden who said that they wasted their money, or the Swedish government's money.

I think that the demise of the industry in Britain was inevitable. In 1977, just before nationalisation, when they were trying to put together a team, it was made clear to me that if I wanted the job of finance member of the board then I could have it. I said that I would take it if Tom McIvor [*Managing Director of the Swan Hunter Group*], for whom I had a very high regard, had been made the number one. Even then, I hardly knew a single person in a senior position in British Shipbuilders. They brought in Tony Griffin [*as Chairman of British Shipbuilders*], who is an admirable Admiral and nobody's fool, but he had been a Royal Navy man who had just recently come into the industry. He had no business experience as such, so I do not know if he was the right chap or not. There is no doubt that, in my opinion, the industry was doomed.

On the question of whether management abdicated its responsibility to union power, how would you have dealt with a lot of fellows who were either quarrelling among themselves over demarcation or quarrelling with you to get more money, and at the same time to a ship owner who has got a charter for his ship, provided she comes on stream at a certain date? The problem was that you had to keep your suppliers toeing the line, the unions and your customer. Each of them was in a position to hamstring you.

Interview No. 21: Eric Crowdy, Managing Director, Hawthorn Leslie, Engineers, George Clark NEM, and Doxford Engineering

I had always been interested in marine matters. I did two years at Cambridge during the war which culminated in a wartime degree. After the war, I went to the University Appointments Board, and the then Managing Director of Hawthorn Leslie Engineers, Sir Philip Johnson, asked me if I would be interested in joining them, so I did. I had various appointments at Hawthorn Leslie Engineers, who were a subsidiary company of Hawthorn Leslie Shipbuilders. I was ultimately appointed Managing Director in 1972.

Post 1972, we traded profitably for some years until it became fairly apparent that, quite apart from the threat of nationalisation, there was a radical change coming into the industry. At Hawthorn Leslie we looked at ways of strengthening our position. We tried to go into a merger with Smith's Dock on the Tees. That foundered because the brilliant financial characters did not realise that our shares were ten shillings each, instead of a pound, and this obviously changed all the valuations on which the negotiations were conducted, and completely aborted the whole affair at a very late stage. Around the time we also decided that manufacturing marine main engines under licence was extremely difficult. The only feature you could compete on was price. One could not claim any design advantage. We had licences at that time from sundry firms, including Doxford [*the only indigenous designer and manufacturer of marine slow-speed diesels in the UK*]. We were also a Pametrada [*Parsons Marine Turbine Research and Development Association, Wallsend*] member. We sought to widen our turbine manufacturing capability. We then decided that the only possible scope was to get more involved in design so that we could offer something different to the standard products of the licensors. To cut a long story short, we got into cahoots with Doxford to design the Seahorse engine. That was a medium speed diesel main engine, which at the time was in a class by itself. It took longer to develop that it should have done, as we did not always agree. We wanted to break into the power generation market and this was the driving force behind Seahorse.

The Seahorse engine project proceeded until one or two events altered developments. The principle one was that Court Line took over Doxford, and not very long after [*1974*] went into liquidation. So, that led to one side of the funding drying up. The Government, who had blown hot and cold on the project, also changed. The Labour Government, who was in power when we started, was extremely encouraging and helpful. They were then defeated by the Conservatives in 1970. We then looked for funding from the National Research Development Council and at one stage they asked us how much we wanted and when. It was the time that the government changed and it just disappeared like snow of a dyke. So, to cut a long story very short, the development actually ceased because of lack of funds. Towards the end of limited funding from government, they were demanding performance criteria, which was gained from taking the best performance characteristic form every international engine known. We had to beat them all, and to the government representative's absolute amazement we took the best from all the competition and we beat every single one of them. Then they withdrew the funding. We could have sold the Seahorse engine vigorously abroad, especially to the power generating market.

Then, the threat of nationalisation became ever more real. I think we were one of the few firms who made a profit. Then, the situation deteriorated rapidly in that the shipyards were having an enormous struggle to obtain orders, and, due to the advent of the Intervention Fund, in the time of British Shipbuilders, the shipyards were not allowed to make a loss over and above what had been agreed; the way round that, was for the engine builders to take a loss. This was never fully understood by British Shipbuilders, but was condoned by them by default. Because we asked the basis on which we could quote sister organisations within British Shipbuilders, the ground rules were that we should quote as if we had a full order book. We, the managing directors, knew full well that a full order book was improbable in the extreme. In effect, we were quoting at an inbuilt loss. This went on until Robert Atkinson became Chairman of British Shipbuilders n 1980. He could not understand why the main engine building companies had been making losses. If we had not gone along with that arrangement, hoping and being led to believe that there would be an upturn in the market, the result would have been that the redundancies would have been much earlier that they were. Whether it was a wise policy or not, I am not sure; it certainly was not thought through properly. There was no cohesive policy in British Shipbuilders for inter-company trading.

David Stables got into trouble at Doxford with deliveries and prices with the little engine, the Doxford 58JS3. They had undertaken to deliver twenty engines inside a year when the engine was not designed, let alone tried and tested. They got into very serious difficulties with delivery problems, which caused immense problems with the shipyards, especially Appledore. This resulted in Stables being removed, and John Murray, who was then Managing Director of George Clark North Eastern Marine [*itself the product of a merger of two marine main engine-building firms, Clark and NEM*] being transferred to Doxford as company doctor to sort them out. I was then made Managing Director of George Clark NEM. I was slowly going through the rationalisation of those two units in an organised way when further moves took place, although, in all honesty, I can not remember the origin of all of them. When Atkinson was appointed Chairman of British Shipbuilders, one of his very early actions was to get rid of Leo Curran, the Managing Director responsible for engine building. As a consequence of the imminent reorganisation, I was swapped with John Murray. He went back to Clark NEM, and I went to Doxford with the view of Robert Atkinson [*a marine engineer*] that I should reactivate British Shipbuilders engine design capability. After a brief period, this started very vigorously with the engagement of the Scandinavian design engineers, the rump of the Copenhagen-based, Burmeister and Wain. In conjunction with them, we got on with the design known as BS42/100, which profited a lot from the experience and reconsideration of the Seahorse effort. It was a true medium speed marine main engine. But when Atkinson's term of office was not renewed, all funding for the development was cut off. Shortly after that, under his successor, Graham Day, the whole enterprise went into rapid closure. So I finished in early 1984, taking early retirement.

Looking at the major strengths and weaknesses of marine main engine building firms from the 1950s to the 1980s, the weakness was the absence of in house design, and therefore reliance on Doxford, who had not kept up with design and development made by other organisation [*for example, Sulzer of Winterthur, Switzerland, MAN of Augsburg, Germany and Burmeister and Wain of Copenhagen*] This meant at Hawthorn Leslie that we relied more and more on our Licence to build Sulzer main engines. Sulzer had a large number of licensees worldwide, and although there was not much trading between countries, there was no doubt that the Japanese were producing engines much cheaper than the European or British product. As far as Britain was concerned we were competing with half a dozen other licensees. Again, the only ground on which you competed was price, and you kept getting orders if your delivery and quality were good. Our quality was always excellent, and our deliveries were better than practically everyone else, so we managed to struggle on, but at marginal profits. We were always beaten down by those people who were prepared to offer lower prices and quicker deliveries, although they did not necessarily achieve them. We saw that as a great disadvantage from the engineer's point of view. The decline of Doxford was particularly important. I do not think that was appreciated enough. Doxford, at one time, had twenty-seven percent of the world market for marine slow speed diesels. That then steadily declined, and Doxford did not take any effective steps to reverse that trend. When the Japanese build-up started, they sought a Doxford licence, but J. Ramsay Gebbie, who was then Chairman at Doxford; would not allow them to have one. That was an absolutely fatal step, an appalling decision. If Doxford had done that then they would had have got the licensor income of millions of pounds from Japan. They did not, and furthermore, they did not develop their engine satisfactorily. Owing to licensee dissatisfaction, Doxford appointed

Percy Jackson from Mirrlees, but, in my personal opinion he was not equal to, or was not allowed to do the task that was necessary.

Total dependence on the UK shipbuilding market was patently a weakness, and we could not break out of that owing to our position as licensees of continental engine building designers and manufacturers. We depended on the UK shipbuilding industry, and we depended on having a product which we could sell to them. Very few engines were imported. By and large, British owners were loyal to British engine builders or had a close relationship with one particular builder. There were only two engine builders who were independent as nearly all were linked to a shipyard. [*Clark NEM and Kincaid*]

The unions were as bad as management allowed them to be. You could also say, however, that the management were not good managers because the unions would not let them manage. Ultimately, nationalisation was the final nail in the coffin of mangers' attempts to manage. The unions saw it as a meal ticket for life, and it was impossible for a nationalised industry to break out of it. So, the solution was to close the industry down. I think a better solution would have been to break the industry up, to privatise it, but that had not got underway by then. I guess that the view was that the shipyards were not privatisation ready. They certainly would have been if they were warship building firms, but too much damage was done by that time. Damage was done not only to Britain, but also the situation internationally. We suffered from having led in the Industrial Revolution and we had inherited all the bad practices over generations, and could not get rid of them. The union officials acknowledged that if we were a Greenfield industry they would agree to changes, but they could not agree in the shipbuilding industry. They were totally trapped, because there were so many unions involved, no particular union could or did take the initiative. There was an inbuilt attitude among the workforce that if they worked hard then they were letting their side down, and ultimately they were doing themselves out of a job. This was an attitude that was very difficult to deal with at the time. It was noticeably relevant when one had a small squad taken out of the works to do a job away from base. The difference in effort was incredible, but the moment they got back into the works, they felt that big brother was watching them. I used to graph the electricity load in the works against time, and it would creep up from zero at starting time, seven-thirty am, and peak at eleven am, and drop off for lunch. It would creep up again in the afternoon and drop of more quickly in the late afternoon. If electricity consumption was a measure of output, it indicated that output was sixty percent of what it could have been.

Another aspect was the method of payment. At Hawthorn Leslie we had piecework schemes. You would have thought this would have been an incentive to put up the earnings, but there seemed to be a belief that you should not put up the earnings above a certain level or the piecework rates would be cut by management. This was not true as the rates were never cut.

Also the slow rate of work during the standard eight-hour day was a manoeuvre to get overtime working. The workforce were always very conscious of delivery dates, and knew if they held back there would be major overtime work to be done. This was particularly the case in the outfit department. They were appalling in delaying the completion of the work until they were given overtime. You could have said that it was bad management, but it was one man against fifty.

On shipbuilding, the Geddes policy was not followed through. Hawthorn Leslie Shipbuilders was the one and only firm coerced into a merger with Swan Hunter. There were other people who had merged, for example, after the Smith's Dock debacle with Hawthorn Leslie; Smith's merged voluntarily with Swan Hunter. The people down at the mouth of the Tyne, Redheads, got into cahoots with Swan Hunter voluntarily, but Hawthorn did not want to get into cahoots with anyone. They saw themselves as a profitable firm outside, but the Geddes rule said that you could not be an approved firm for subsidy [grants and loans] unless you agreed with their ground rules. Hawthorn Leslie was the most successful shipbuilding company at the time, and they were coerced into bed with Swan Hunter. As a result, their skills and specialities were diluted to nothing.

Some grouping should have taken place to give us competitive competence against Japanese competition, but the way in which it was carried out was totally inept. It was the government who said that you were or were not an approved yard. The only yard that was not approved was the most efficient and profitable at the time. This was not an equitable and sensible management tool. This was due to a lack of understanding as to what made the industry tick.

I am a great believer in free enterprise. I think it is the one thing that the Tory Party [*Conservatives*] got right. For two to three years before nationalisation all the shipyards were thinking of was the best deal they could make under what appeared to be the inevitable. Prior to that, I think it was necessary for some government support; otherwise the reduction of the industry would possibly have been too radical. I am uncertain as to how that should have been administered. Nationalisation, when it came, was a disaster. It could have been much better than it was. The changes of chairmen were definitely significant. I gather that the first chairman appointed was done so by mistake. Having said that I have the greatest regard for [*Admiral*] Tony Griffin as an individual, but he had no experience of the in-fighting of a commercial organisation of that type. He had a diabolical Chief Executive, who was not recommended to succeed him. In my book, the Chief Executive, Casey, was not a suitable person for the job. Then there was a long period of no one at the helm, before they appointed Robert Atkinson, who, if he had been allowed to continue, might have done something successful. One of the worse things that happened was that they allowed, due to government interference, the only young, knowledgeable and competent shipbuilder to be taken away from them and sent to

Harland and Wolff. That was John Parker. The rest were either old, or wrapped up in family dynasties. Atkinson was sacked because he would not privatise the warship builders. It was totally irrational to privatise the warship building industry whose only customer was the government. [*They rationalized it on the grounds that it would induce greater competition among the firms, and lead to a better deal for the taxpayers*]. Then, Graham Day was the man who was installed to dismantle it.

As I saw it, the government saw British Shipbuilders from the Treasury point of view as a source of undesirable expenditure, which was not commensurate with its employment potential. I think they got their sums all wrong and did not realise the extent to which the shipbuilding industry had a supply industry which employed many times more. On the other hand, the decisions were taken at a level that said that this was something we could do without. I do not know whether this was right or wrong. Japan, obviously, took the opposite decision, as did Korea. Perhaps, as an island nation we were too ready to adopt that approach. Other nations had managed to retain a shipbuilding industry, notably Finland.

To some extent, the survival and failure of the industry was wrapped up with Government. They did not subsidise British ship owners the way the Americans did. We decided not to do that because it was against the General Agreement on Tariffs and Trade, and those of the European Free Trade Area, which was also hurled at us by successive governments as a handy excuse. Every other country broke it except us. I do not think they appreciated the need for a British Merchant Marine, or the Merchant Marine's unique position in cross-trading. If the Falklands War had come a bit earlier they might have sat up and thought a bit. I think that Margaret Thatcher is a total land-lubber. Her Channel Tunnel mentality typifies her thinking. She personally had a lot to do with what happened. I do not think it would ever have happened under Labour politicians, but they might have got us into worse difficulties.

Interview No. 22: M.F. Pyman, Managing Director, Mercantile Dry Dock Company, Jarrow

I was lucky enough to leave school in the middle of the war when the authorities wanted technical people. I was good at maths, so I was sent off to Cambridge to do an engineering degree with war service to follow. I chose the Navy, and became a Midshipman, and left as a sub Lieutenant in 1947. Instead of going back to Cambridge to complete my degree, I went back to school at the St. Peter's Engine Works of Hawthorn Leslie for a couple of years as an apprentice. I also spent a bit of time at Doxford, and at sea and came back to the Tyne with the Mercantile Dry Dock Company at Jarrow around 1953, became Managing Director, and stayed there until nationalisation.

The Mercantile was a fairly small yard in the 1950s, but after we built it up we ended up with four docks. The job was purely ship repairing. We were taken over once or twice prior to nationalisation. We went in with North East Coast Ship Repairers, and then that, in turn, was taken over by Court Line. When Court Line collapsed, we were taken over by the Government by Tony Benn who was Minister for Industry at that time. When nationalisation came we were in with British Shipbuilders, although not all the repairers were.

The strength of Mercantile, and ship repairing on the North East Coast generally, was the traditional expertise of the men. We had a huge pool of skilled labour at the time, every sort of sub-contractor you could think of could be called in at short notice, and we had very experienced middle management. We were also better placed than other districts on the working practices side of things. Also, in the 1950s we were well placed geographically with timber [*Baltic trade*] and coal supplies. This changed dramatically in the 1960s when we started to really struggle with the timber finished and the coal virtually finished. We had just built a new dry dock, so we were really struggling under financial constraints thereafter.

The weakness after 1960 was this same geographical position. With virtually every ship we got we had to attract them to the Tyne, so our prices had to be that much lower. In the 1970s when the price of oil rocketed, it became very difficult to attract ships to the Tyne.

The small scale of many ship repairing firms was a positive in many ways. It kept overheads down, and we did not need too much purchasing power. Lack of investment was no problem. If anything, we over-invested. Was there obsolete technology? In ship repairing a lot was obsolete, but it was very difficult to do it in any other way. Some of the top management, me included, did not have good all round ability. I was trained as an engineer, and had to pick up a bit of accountancy, a bit of industrial relations. Generally, that was the case for most firms where the boss had training in an engineering background. Labour relations were not as good as they should have been, but they were not as bad as in some places. Individuals professionally trained in industrial relations did not really start to come into the industry until the 1970s. I usually did it myself, which meant that it took up a lot of my time.

In ship repairing, we would claim that productivity in steelwork was particularly good, and was quick mainly due to the expertise of the labour force. The fitters and engine work was not so good, mainly due to the proliferation of trades. Lifting the rudder is the obvious example where you had five trades involved. One had to wait for the other for a very simple operation. We never got that problem right in my time. It had been going on for so long, and the men had got it running so slickly that we as management thought the situation was quite good.

Marketing was something we should have done more of in ship repairing. I did the marketing myself, although we did have agents in London and abroad, but it was not really a planned operation. A lot of it was personal contact, and you tended to concentrate on those owners with whom you had dealt with in the past instead of going out looking for new horizons. As to what we could have done about it. We were small, and probably a marketing department was beyond us, but we could have used the agent system more effectively.

In 1959 we were looking for new markets for our new dry dock, which would take 20,000 ton ships because the market for colliers, and timber ships from Scandinavia was declining, as both trades were closely associated with the fortunes of the mining industry. It was difficult breaking into the big tanker companies for a small yard which had not done much tanker work. Around 1962, we started on the issues of flexibility and interchangeability of trades, the key words in the Patton Report on Productivity in Shipbuilding. We got a little here and there, having bought the changes, but not enough. We would raise the issues around the time of the annual wage negotiations. The Boilermakers in particular were difficult. Part of the problem was your own best men, the 'royals' as they were called, the platers who had a lifetime's experience. They were not going to have any interchangeability, and the foremen rather backed them up. The biggest problem I had, however, was communications. We were not taught to communicate properly. I used to leave it to the shop stewards to try and convince the men. It was very difficult to crack the hierarchy.

In terms of management I had three men at the top; the Company Secretary, who ran the office, the Chief Estimator, who was very important and settled all the accounts, and the yard manager. I tried to leave them all alone as far as was possible to do the work. Under the yard manager, were three or four ship managers who would have charge of the ship whatever the size of the ship. They would usually be from an engineering background; sea going men who would have got a Chief's ticket then that was one up to them. However, this was not really applicable once they came ashore. Some of the guys who only had a second's ticket were probably better than the others. The parameters for picking a man were his background at sea, which I do not think was right. It was the same thing on the steel side. You would pick someone with drawing office experience. He would probably have got his Higher National Certificate and had worked in a shipyard. We used to shamelessly pinch people, of whom there were plenty, from other Tyne yards. They were not retrained for ship repairing. They had to learn from experience.

There was a big development in offshore dredging, and that became quite big business, so we tried to capture a bit of the market. Then, when the offshore oil industry came along in the North Sea, supply ships were an immediate source of repair work for us. They were smaller ships which is all we could take at the time. In terms of competition, our main competitors were in Rotterdam. The Dutch and Scandinavian yards were always competitive, but we did not bother too much with France. The Americans kept o their own country by and large with a fifty percent enticement to their own yards for repair. We spent a lot of time in Scandinavia, and had agents in Norway and in Finland. The Greeks were always there, an amazing variety of London Greeks, some of whom were excellent ship owners and very good to work for. We had a wide variety of customers. However, I never found the competition unfair. The Dutch yards had an advantage in that if a ship owner took a ship into Rotterdam for repairs he did not pay any port dues, which was a continual niggle with our Tyne Port Authority. The various charges on the Tyne made it almost impossible to have a two-day docking, and a quick wash and brush up on a ship.

I do not think that nationalisation was necessary for the ship repair industry. I think that when the Labour Party Manifesto writers saw that there were some ship repair yards within shipbuilding yards, they would just chuck in the lot. If you nationalised the car industry, would you chuck in the repair garages? Under nationalisation, we continued as much as it had always done. It was a big change for me as a new boss came in immediately. I was made Marketing Director, which I was not all that good at, but I accepted it on the basis that if I did not do it, they would probably get someone else in who was even worse. There was quite a change round of senior personnel.

Nationalisation gave the advantage of enormous buying power for raw materials we needed, and, theoretically, a great pool of expertise on any subject you might mention. On the other hand, you had to pay for it in overheads. This was payment for often over elaborate structures and offices. The other disadvantage was that nationalisation was a bit of dead hand. You were allowed to travel first class everywhere, which was not in my view, necessary.

Probably the biggest blow to ship repairing on the North East Coast was the increase in the container ship, and the ferry traffic over the channel. Most of the cargoes that used to come in ships now go by rail to Dover. Ships were going for longer and longer period without the need to dry dock. It used to be every nine months, now it is every three or four years.

In shipbuilding there were amalgamations which were not properly understood by some of the old bosses, many of whom were family descendants of owners. Some of the yards who did not have a good reputation, or were geographically badly placed, or did not have a good workforce should have been shut before they were. Everything should have been concentrated where it mattered most, instead of trying to keep the old practices going, on the management side as well as the workforce.

Interviews
The Wear

Interview No. 23: Arthur Storey, Director North Eastern Marine, Director Doxford Engineers, Director Vickers Engineering, Barrow; and Managing Director, Austin and Pickersgill

In September 1920, I went to work at NEM. I was with them for thirty-two years. In 1953, I went to Doxford to take over as Director and General Manager of their Engine Works. I stayed there for five years until joining Vickers at Barrow to take charge of their engineering side at a time when they were just starting on their nuclear submarines. I left Barrow in 1968, and joined Austin and Pickersgill as a non executive director. When the then Chairman, Charles Longbottom, and the General Manager, Jim Gilfillan left to join Court Line, I took over as Managing Director until I got Derek Kimber to join as Chairman and Managing Director. I carried on at Austin and Pickersgill until 1976, where I assisted in the formation of A. and P. Appledore, and have been associated with them ever since.

When I went to Doxford, we had huge marketing potential for the Doxford engine. I put in an application for the extension of Doxford's engine building capacity, but it was vetoed by Dr. J. Ramsay Gebbie as soon as Dr. Andre, the Chairman at the time who had always solidly backed me as director, retired. Doxford engines were the best, and in the period after the war it was out selling the Sulzer engine. If I had been allowed to develop the engine as I wanted to and to extend production of it, it could have remained the best selling engine. But, Dr. J. Ramsay Gebbie put the block on it in 1957. He told me that I could forget about extending the Engine Works because Doxford built ships. However, the shipyard could only produce six ships per year. With the market values at that time, the maximum profit the shipyard could possibly make was about £350,000. Whereas in 1957, I turned in a profit of about £1.6 million on engines, we were producing eighteen engines per year, and they could only build six ships. So the decision to halt development and innovation on Doxford engines was down to Dr. J. Ramsay Gebbie. This can not be overemphasised. You can find my statements to that effect in The Transactions of the North East Coast Institution of Engineers and Shipbuilders. This was very important for the future of marine engine building in Britain.

I had excellent industrial relations in both Vickers and Austin and Pickersgill. In Barrow they went on strike, and I spoke to the union people over one weekend, and persuaded them to come back to work. I telephoned

Len Redshaw, the Shipyard Manager, who was very angry, asking me why I had bothered to talk to them. At the meeting next day you could see his attitude. He arrived late, and told the men that if they wanted to go on strike then they should just go ahead. His approach to labour relations was too dictatorial really. My personal relations with the unions were very good, when I was pushed out of Barrow; I got a beautiful letter from the Confederation of Shipbuilding and Engineering Unions. If management treated unions as partners and not employer and employee, and had regular meetings with them then industrial relations problems could be resolved. I do not think management abdicated its responsibility to union power, but there was not the give and take that there should have been. I can not recall any yard where the unions took control. Senior managers were not very well trained. I had a policy in the yards were I worked, that you selected the promising youngsters and set about grooming them for management, sending them all round the world, and training them to get orders.

Graduates were a mixed blessing. I preferred the up-grading of shop floor people who had started as apprentices. I had a policy of moving apprentices around the yard so that they got as wide a conception of all the various aspects of production. Then I would bring them up into junior management. It was a very definite policy which was not always followed by other firms. Some, like Swan Hunter, preferred to bring in graduates. Graduates were coming in with an academic intelligence, but could not really match the knowledge and skill of apprentices who had been there from fourteen years of age with seven years all round work experience.

The levels of education and training were higher in Japan. I recall a visit to Mitsubishi. When I asked them about their methods of training, they told me that they had learnt it from the British yards in the sense that they practised what they preached. We were not activating the thinking as much as the Japanese did. They trained their youth, and built up a formidable position. As well as training their people better Japanese yards worked better. There was a great family tradition in Japanese shipyards and engine works, and they looked after their employees from the cradle to the grave.

World slumps affected you, but against a world slump you have to develop your management strength, and hold that. Whatever the world situation was you had to have men active in the marketplace to pick up orders. You had to spend money in the best production facilities you could create in a yard. Investment was always very high on my list of priorities, but unfortunately that was not always the case in many yards. Lack of investment was a major reason for the industry's decline, but not the only one. On the Wear there was little investment at Bartrams, none at all in Short Brothers of any note. In Doxford, investment was strong in the Engine Works. Bartrams eventually merged with Austin and Pickersgill, but they had a very weak management. The takeover was a natural progression. It was very painful to watch the yards on the Wear wither and die due to poor management and marketing ability.

At Austin and Pickersgill, marketing was really the secret, and the great thing was to get owners interested in buying a standard ship, and we really went to town on the SD14, that caught on very strongly, and we built about 180 of these standard ships. I went out and brought owners from all over the world to the Wear, particularly from Greece, and I also brought in owners from Argentina, Japan and Holland.

I had mixed feelings about nationalisation. I did not see it as being merely a bad thing because, obviously, the yards were not doing as much for themselves as they should have done. Nationalisation could have brought about more rapid change and improvements in yard facilities. The merging of yards was quite a good thing in principle, but the bad thing was that the Government did not have within its core top quality industrial advisors.

Looking at the industry overall during my time, one thing that really sticks out for me was the general nepotism in the industry. That was one of the things that helped to start the rot. Management was not developed in a sensible manner. Often members of a family were brought in who could not care less. They had a family shipyard, which they just milked really. The lack of investment in yards owned by families was very pronounced. Nationalism went a big way to get the families out, and eventually they did get out. When I went to Sunderland at first in 1927, there were eleven or twelve shipyards, and latterly there was not one single member of those families left in business. The last one was Jimmy Marr at Laing's. Nepotism and a lack of strong marketing were the two key factors in the decline of the industry.

Interview No. 24: Ken Douglas, Shipyard Manager Vickers, Tyne, Director and General Manager William Gray, Hartlepool, Managing Director, Austin and Pickersgill, Sunderland, Managing Director UCS, Marketing Director, British Shipbuilders Repair Division, and again, Managing Director of Austin and Pickersgill

I left school in 1936, which was just coming up to the end of the 1930s depression in which my father and grandfather, both shipyard workers, had walked the streets for about six years. I began an apprenticeship at Laing's in the drawing office, and started to study at evening classes, took a scholarship to Sunderland Technical College and qualified in naval architecture. I then spent eighteen months within the Ship Division at the National Physical Laboratory, and left to go to Vickers Naval yard on the Tyne. I spent my first fourteen months there setting up a work school for apprentices, and then went into management and building ships. I spent eight years at Vickers building passenger liners, warships and tankers. In 1953 or 1954 I was invited to become Director and General Manager at William Gray at Hartlepool, ran their two shipyards, and handled their repair side and labour matters for five years, before leaving to become Managing Director at Austin and Pickersgill at the age of thirty-seven in 1958. I left Sunderland in 1969 to go to UCS and when it went into liquidation, I agreed to carry on as Managing Director with the Liquidator, Robert Smith, and stayed there for a further two and a half years. In 1978 I was persuaded to join British Shipbuilders as Marketing Director Repair, and a year later I returned to Austin and Pickersgill as Managing Director, lasted five years, and retired.

When I came back to Sunderland in 1958, as Managing Director of Austin and Pickersgill [A&P] some modernisation had begun, but the Board took the water in, to use a euphemism, and the managing director at the time said that they either went ahead or they agreed to find a purchaser and get out. So the shareholding in the company was sold to Basil Mavroleon [*of London and Overseas Freighters, the largest independent British tanker company by 1960*] and we went ahead and finished the modernisation. I gave myself two years, despite the fact that the chairman said I was barmy and that the place was bankrupt. As things went, we finished the modernisation and were building ships very quickly with the most modern shipyards in Europe at that stage. We had no strikes for eleven years, very good labour relations and a very good labour force that were well paid, and produced a very high quality product. At the time I left to go to UCS in 1969, there was only one Scandinavian yard competing with us on the level of steel throughput.

At A&P we had gone through a process of getting rid of individual ships and had introduced a policy of building vessels to a standard design, with an interest in bulk carriers of 16,000 to 30,000 tons deadweight of which we built half a million tons deadweight. Seeing the market was going to go down, and looking to what was coming next we thought that the Liberty Ship could not go on much longer, so we developed a ship which became the SD14. At the time I went to go to the Clyde there were about forty-five on order. We were competing with the Japanese. This was before the subsidies were introduced. They were selling their Freedom Ship for £1 million, and making a loss. So the ability was there on the Wear to compete, and so to, was the competence of the workforce.

One of the strengths of A&P was as a result of the scheme of modernisation, which consisted of raising the shipyard to the ground and rebuilding it completely as a modern yard. There was none of the old yard left by the time we had finished, and we continued to build in our East yard during modernisation, thus we had a stepped changeover programme. By the time we launched our last ship from the East yard, we were in full programme building all-welded ships in the West yard. By 1961, we had the most modern yard in Europe. By that stage, whereas in the past ships would have taken two years to build, we were now building 18,000 ton deadweight vessels in 102-105 days, which would have been inconceivable under the old system. Because of the new concept, and a young team, we had strength in that we had something new, we had something to show people, and there was a tremendous enthusiasm which we got across to the workforce. Also, I could not accept the extant system of payment; I wanted incentive, output bonuses etc. We got these things in a very short period of time. The only strike I had lasted for five weeks, was because when I had meetings with the men, I told them that I would not be satisfied until I saw them going out at the end of each week with £20 in their wage packets, they thought there was some catch in it, and went out on strike. Eventually, we negotiated and got the incentive schemes going; various tonnage schemes and bonus schemes. This worked very successfully. By the time I went to the Clyde in 1969, the key men in A&P earned over £40 per week, whereas on the Clyde, the men were on a low rate of £23 per week.

So I think at A&P the main strength was good labour relations. I used to talk regularly to the men, and not just the shop stewards, and let them know what was going on in the yard. We had a good team going, and everyone was enthusiastic, but with the economic circumstances of the time in the mid 1960s we could not make big profits. I think another strength was our security of work. We got the weakness out of the system, a weakness which existed in British yards, of trying to please their owner friends and pandering to their quirks. Sweden [*a neutral country*] had a very good war. They were able, having spent time at Vickers Naval yard before the war on welded ship construction, to go back and develop the techniques, modernise the yards, and install large cranes, pre-fabrication shops and so on. The yards in Germany and in France had been severely damaged during the war. We had a situation in Britain at the end of the war where we were not able to modernise because we had been too busy turning ships out during the war. Then because of a few disasters with welded Liberty Ships, owners were very conservative, and now that the war was over, they insisted on getting back to riveted ships, for which the yards were equipped. On the continent, they started wholesale on all-welded ships, and then we had owners coming along and telling us that they had a similar ship built in Sweden with 500-600 more tons deadweight. [*Welded construction facilitated greater deadweight carrying capacity and a more streamlined ship than did riveting*] I got out of the system by simply insisting on standardisation and telling owners to go elsewhere if they wanted other things. Having been put in that situation, they then looked at the prices of others, and we were very competitive. I was selling bulk carriers on an across the counter basis, £55 per deadweight ton. So we overcame the weakness for a demand for an increasing variety of ship in what was a restricted resource.

This was the fault throughout the industry, because as all the types of ships developed after the war we had yards like John Brown building a variety of ships for which they were not equipped. At A&P we had got that weakness out of our system. We had got the economics and the concept of the standard ship over to owners. By the time we had finished our bulk carrier programme in about 1967, we were then into the SD14 programme, which went on to build 126 on the Wear, and 237 worldwide.

When I went to UCS in 1969, the obvious weakness was that the initial investigation into the group's viability did not reveal the facts. Post Geddes, they grouped five yards together, which, left to their own devices, would have probably gone bankrupt in any event. So grouping saved them for a time from bankruptcy, but there were inherited losses in the programme for the five yards of £22 million. On that basis, what was grouped was really a limited company without any capital. It was depending on the Government for survival. At the same time, there was a resource there. When I went there Stephen had been closed. They had the best preparation and assembly shops of all five yards. I reopened these and started producing fabricated steel for the four remaining yards. We fabricated steel in Stephen on quite a big scale.

I think at UCS there was a tendency not to realise what the modern world was all about, and to try to cater for all tastes. The resources simply were not there. John Brown was never a big shipyard; it was two medium sized yards, the East yard had big berths capable of building big ships, but they did not have big steel production facilities. In big passenger ships and warships, the steel content is only a very small part of the ship. In so far as there is a real problem in the construction of these types, it is the logistics of getting the material and equipment that you need at the right time, and in the right place. What you had at John Brown was an absolutely magnificent production shop for joinery work. It was really a fitting out yard, but the West yard was not as big or as well equipped as A&P at Sunderland. The point is, that I had been turning out with 400 men about 500 tons a week at Sunderland, whereas at no time during the construction of the *QE2* did they need to produce more than 125 tons of steel a week because the vessel, obviously, was much more sophisticated. So John Brown was not a good steel producing yard. They produced good steel, but it was not good for producing ships where steel was the highest content of the ship. Steel throughput was slow, and it was never really set up for mass production of steel.

In fact I thought I was going to have a civil war on my hands when to get more space in the preparation shops in the East yard, I gave orders to take the old wall-mounted drills off, which had been used to drill armoured plates during the First World War. These drills and the beds were looked on with great pride; they were sacrosanct. This was like the furore I caused when I insisted on managers wearing safety helmets rather than the traditional bowler hats.

Eventually, I had to say that for the future development of UCS, John Brown would have to go. I had fierce arguments with Jimmy Reid about it, although he was always a friend. He was concerned about the contribution to Clydebank that the industry made, but Connell was quite a good yard which produced very well, Stephen too. The weaknesses in the Govan shipyard [*Fair-*

field] tended to be with logistics. I found the steel throughput very low. During the previous decade prior to my arrival, Fairfield steel throughput had averaged 265 tons per week, and that was with 1600 steelworkers. I had just left a similar situation with 400 men producing 500 tons per week. The press blamed the men, but really it comes back to blaming management. I would go to the assembly shop, and there would be three plates missing so they could not finish the job. They were not insisting on the steel mills sending the steel in the right sequence. I saw the steel mill owners and told them if they could not get their act together, then I would go elsewhere.

As well as that, we were on an agreed redundancy programme, which was properly negotiated with the unions, and we were steadily running down to an economic level in manpower. In about fourteen months, I had reduced the workforce from 14,000 workers down to 8500 workers without any strikes. The unions fully accepted that, the big ships having gone, that the resource was only there to produce with an £8m a year wage bill, which previously had been £13.5 millon. They appreciated that, and we worked together on it and I do not think the men suffered. As far as Government was concerned, with getting the Stephen shops back into action, we had reduced from 1600 steelworkers to 1100 steel workers, and we were producing 600 tons of steel per week as against a previous average of 265 tons. There was thus a two and a half time's increase in productivity.

The problem with UCS was that it was under-capitalised. It started off in debt, and there was no way it could get out of debt. There was an idea that with getting rid of what they thought was the old fashioned shipyard management and bringing in newcomers from other industries, who knew nothing about shipbuilding, that this was going to be the answer. But, it was not; the productivity was not there, people had not got down to the simple things like getting the steel deliveries right. It was not the fault of the men but the management for not getting the supplies on time. There was not sufficient dynamism being put into chasing up suppliers. Management had possibly become a bit punch drunk with being hammered with the argument about late deliveries, and they were striving on with the best will in the world, but they were not looking at why it happened. There was a shortage of original thought there.

UCS tried to standardise on design and methods, which was absolutely impossible, because all the yards were differently equipped. They had different sized cranes. If you design a ship with on average fifty ton lifts for pre-fabricated steel work, and one of your yards had only twenty ton cranes, then you had to alter the drawings and split your units to suit that yard. They did not all have the same balance and facilities, and there was no way you could talk about a common design office and common methods with such different facilities. What should have been done, in my view, and which was to be taken up later, was that grouping should have been done in the form of a holding company. The individual yards should have remained as limited companies, standing on their own feet, and answerable and accountable to the holding company. What happened is that they put it together as a conglomerate. When I looked around and decided that the John Brown yard had to go, and as the finances were all in one pot, I could not close one yard without the Bank foreclosing on the lot. Say the closure of John Brown cost £3m, but it would not bring in £3m on liquidation, then the Bank would come and tell us to close another yard so that they could recoup their investment. The whole financial structure was put into one pot. They should have learned from the Greek ship owners, like Onassis and Niarchos who made every ship a limited company, so that when one went bust it did not interfere with the rest of the fleet. Therefore, on the Upper Clyde, you could not say that this limb has to be cut off, because a bank would come along and demand the arm in compensation, even though it was not gangrenous.

Eventually, I talked this out with the Conservative Government's Minister for Industry, John Davies who accepted my plan of survival to offload John Brown, and also to take all the land between Stephen and Fairfield to put down two large berths to build ships up to 150,000 deadweight tons, make it all into a modern complex, and in the meantime, because of the social side of the workforce, to keep Connell going on the other side of the river. In that modernisation, which was similar to what I had done at A&P, but on a much larger scale, we would have finished up with the most modern yard in Europe. I then wanted to bring the workforce over from Clydebank, phase it in as the modern yard developed, and eventually close Connell down, and have this big complex at Stephen and Fairfield thrown into one. Davies was prepared to buy that. It would have cost about £12 million plus a capital restructure, which meant the Government agreeing to write off the £11 million in loans that we had. He agreed to that. Eventually, though, and I will not go into the reasons why, the decision was taken to close. It would make your hair stand on end. I had a talk with John Davies who told me that the Government were not prepared to put in any more money. I knew what the implications were. A number of owners had placed orders with us because of old personal connections. I called a meeting of the ship owners and put a number of points to them. All twenty owners who attended agreed to six percent increase in prices. This gave us the £3 million shortfall. I then called Davies the same day, I told him that we had the money, the £3m shortfall. They were determined to find a way to collapse us, there was no question about it. By that time, we were starting to make a profit with levels of productivity rising, and the labour force reduced.

We had got the mix of ships out of our system, which had been inhibiting production. We were simply down to the standard Clyde ship, and the Cardiff class bulk carrier, originally designed for Reardon Smith, but with a little modification, was in demand by other owners. At the time we went into liquidation, the Order Book was £160 million strong, and we were collapsed for £3million.

On international competition, the question of subsidies is a difficult subject. In my view, subsidy in the long-run never works. The Japanese did not introduce subsidies on the scale that people talked about. They put restraints for nationalistic reasons on Japanese ship owners ordering abroad. For example, a Japanese owner building in Japan could get twelve years credit at 3.5 percent, whereas the best anywhere else was ten years at 5.5 percent. On that basis alone, the Japanese owner would have to have a hole in his head to build abroad. So they are not directly giving money, but it can be translated into what sort of subsidy is needed to compete with these restraints.

At the end of the day, it was not a question of subsidies; it was a question of wage levels. People talked about Korea and Japan and how we had to get productivity up. I tried later on with British Shipbuilders to spell out that Korea was paying their men the equivalent of £35 per week for seventy hours, that is, fifty pence an hour. I was paying my men £3.50 an hour. If productivity was equal, then a £3 an hour differential on a ship with a million and a half man hours meant that we were £4.5 million behind before we started. That goes right through the overheads. So they were, because they had taken from our designs, offering a similar 26,000 ton bulk carrier to us at $15 million, and after eighteen percent subsidy our best price was £15 million. Productivity has nothing to do with it, because if you look at the basic point, suppose productivity is at sixty man-hours per ton in a Japanese yard for preparation, assembly etc, they were paying £30. In a British yard at £3.50 an hour has to produce it for £30, we have got to do it in eight hours. It is impossible to do, so the argument about the lazy workforce, and the inefficient management, and how productivity was so much lower than the Japanese and the Koreans was essentially the difference in base wage rates.

In 1983, there was an international conference in London, and all the shipbuilders were there. We talked to the President of the Korean Shipbuilders Association. We asked him why they did not quote even higher prices, because even If they did, we still would not be able to compete. He said they were not worried about us or any other European competition, or even about Japan, but they were worried about China. China now has twenty shipyards, and at the time a senior executive in a Chinese yard got the equivalent of £40 per month. The fundamental point is that shipbuilding is a labour intensive industry, and the largest element of cost is labour. If there is a large differential in the basic labour rates then there is a productivity gap which can not be closed.

I think that for all European governments shipbuilding had been an embarrassment because it is labour intensive. We are competing to a large extent with low cost third world countries, and there is no real answer. Any sensible government had to see the writing on the wall, and realise that there was no way the taxpayer could be expected to go on slogging his guts out to pay a level of taxes where we are going to need forty percent subsidies just to compete and stay in shipbuilding. In Britain it came down to a political decision by the Government. In my view they should have faced up to it in a more Churchillian way, and put the facts to the country, and not just to the shipbuilders. The Government ought to have said that the industry had to be replaced, as indeed, it is now being replaced in Sunderland, instead of going about piecemeal and shutting this and that yard. If the Government had faced up to this, I think that we would have been a long way further ahead with our more modern industries, which are now being captured by the Americans and Japanese.

Industries have come and gone; America realised this years ago. They really do not have a shipbuilding industry for the purpose that people normally understand, as a commercial proposition. America decided many years ago that she needed to maintain shipyards and ships for defence purposes. They took their shipbuilding subsidies into their Naval Estimates every year, and then operated the differential subsidy system both on freight rates and shipbuilding costs. An American ship owner got the best price he could from an American yard, and from foreign yards the difference was subsidised. The only way, to my mind, that shipbuilding could have been carried on in this country on a similar basis was if the Government wanted to admit that it was required for defence purposes, and take it on the Naval Estimates. Whereas with a commercial proposition acting against other commercial countries, then there was no way that we could carry on with shipbuilding.

There was, however, some room for shipbuilding on a niche basis, and this was a point I made to British Shipbuilders. At A&P by 1983, so far as series production was concerned the SD14 type was going to the Third World producers. We would never get that back. There were, however, still specialised ships we could use the facilities for. In order to keep yards going you had to change the concept to a different market altogether, with specialised ships that nobody else wants to build. I give the example of Fords at Dagenham, if you ask them to build you a Rolls Royce; there is no doubt that with their equipment that they could build one. But do they want to do it in a production line of 250 Escorts?

There is no question in my mind that nationalisation was not necessary. The industry would have survived and would still be surviving with the cyclical nature of shipbuilding, because whilst we had those wage differentials, in time it would all have adjusted. Where nationalisation missed out was that it had to get into bed with the unions to formulate a policy which was acceptable to both sides and to government. In doing that, when things got rough, they went ahead with a no compulsory redundancy agreement; it had to be voluntary redundancy. That meant that they played the numbers game. In my final year they came along and said they wanted 700 more redundancies when I had 2300 men. I got applications from 1000 so there was 700 names picked out of the hat, and they got their lump sum payoffs and were out. The other 400 men who remained in the yard were a cancer. They wanted redundancy so they were not working. This was a running sore in the yards. The one thing that British Shipbuilders did right was leave all the yards as limited companies. I think that they committed the cardinal sin of buying things they did not want with money they did not have.

Interview No. 25: Derek Kimber, Shipyard Manager, Fairfield, Harland and Wolff, Belfast, and Chairman and Managing Director, Austin and Pickersgill, Sunderland

I finished an engineering degree and joined the Royal Corps of Naval Constructors in September 1939, and commenced another degree course as Constructor Lieutenant. I qualified as a Naval Architect and spent ten years with the Navy. I then joined a firm of management consultants for four years, before being headhunted by a friend to join Fairfield as a Shipyard Manager in May 1954. I stayed with Fairfield for twelve years, before I went to Harland and Wolff. I left Harland and Wolff to become Director General of the British Chemical Industries Association, and after three years there I was headhunted by Basil Mavroleon for what I understood to be the job of Managing Director of Austin and Pickersgill, but during my first interview I was offered the job of Chairman and Managing Director. I was there for over ten years, from the beginning of 1973 to the end of 1988. By this time I was sixty-six, and I decided that I had had enough of the management of British Shipbuilders, for whom I had neither time nor much respect. I was very happy to agree with Graham Day that I should retire.

One of the strengths from my point of view at Fairfield and Harland's was the fact that the top people, by which I am referring to the Chief Executive and Chairman in each case, were very flexible in attitude, and gave me a free hand to do what I wanted to do, firstly in rebuilding both shipyards, and secondly, managing them. That did not apply so much from the start, but in A&P, because I was the boss there, I did have a very free hand from Basil Mavroleon, who was Chairman of London and Overseas Freighters who owned the yard at that time. He never interfered, although no one could have been more interested in shipbuilding than he was. The steelwork facilities at Fairfield and Harland's when I first went there were pretty much a load of scrap. Fairfield had not had anything to speak of it done to it since before the war. Being part of the Port Glasgow based Lithgow Group, Sir James Lithgow, when he was Controller of Merchant Shipbuilding and Repair at the Admiralty for the duration of the war, was determined that no one could say that as a result of his key government position that he had rebuilt his yards with taxpayers money. Harland's was a rather desperate amalgamation of four shipyards, and I do not think anyone knew where to begin in modernising the place. So, one almost had a Greenfield site in which to work on in each establishment, and this was strength.

The weakness was that both of these yards had almost run out of money by the time I arrived. We did manage to scratch together about £6m to £7m at Fairfield, and a somewhat larger sum at Belfast. But we were really short of capital from the start. The other major weakness, and the one I would probably put first, applying generally throughout British shipbuilding, and the major cause of us falling off out perch as premier world shipbuilders, can be summed up in one word, tradition. Tradition is a most dangerous word. We thought that because we had been building the world's ships for so many years that we knew all about shipbuilding. We were very, very slow to realise that other people might have different and better ideas, and very, very slow to adapt. Tradition, in a slightly different form, permeated all our trade unions who wanted to carry on doing things exactly the way they had always done them. We had all these trades jealously guarding their various elements at work. Tradition, above all else, brought British shipbuilding to its knees. It was not until the mid 1960s that we started to pull our socks up, but by then it was too late.

When I got to A&P in 1973, we had overcome most, but not all, of the industrial relations problems in the industry. Certainly at Fairfield and to a lesser extent at Harland's where the lads, I thought, were more intelligent than they were in Scotland, we spent a tremendous amount of time sitting across little wooden tables arguing with shop stewards. I think I averaged something like 28 hours a week in my first five years at Fairfield arguing with shop stewards, when I should have been managing the yard. In all I had to deal with twenty-one unions at the Govan yard and there were another clutch in the engine works.

Another strength was that the craftsmen were, and are, really skilled men. I think one of the reasons why in other countries people had to change their methods and change their whole shipbuilding philosophy was that they did not have this large pool of skilled labour who could do the work, so perhaps this strength was also a weakness.

Another perceived strength at the beginning of the post-war period that turned out to be a weakness was that British shipbuilding did not use ship brokers to obtain business to any great extent. I remember James Lenaghan, my boss at Fairfield, being critical of the use of brokers. You have a much larger customer base if you offer your wares through substantial firms of brokers, but of course you still have to offer them at a competitive price. I do not agree that in the 1950s and 1960s that our shipbuilding firms were smaller or substantially smaller than those in the rest of the world. As the 1960s wore on they became so. We had our "big six," John Brown, Fairfield, Harland and Wolff, Vickers, Cammell Laird, and Swan Hunter, and there were similar sized firms in Germany, France and Scandinavia. But by that time there were several yards beginning to spread their wings in Japan. We did show the same enthusiasm for investment. Some people say that was because so many of the yards were family owned, and the families put the profits into their own pockets. I think that is a bit unfair. Obviously the owners made sure that they were not going to starve, but I do not think that they were unduly greedy. They did not make as much money as people think. They did not, however, accept or even perceive that the shipbuilding industry was on the threshold of enormous changes, and it took them too long to accept that they would have to change too, or they would go out of business. By that time they had lost their position in the race. Such changes that did take place occurred too late, and often were not swingeing enough. At Fairfield, we were one of the first firms to introduce a number of new techniques like tenth scale lofting and automatic cutting machines, closely followed by others. However, because we had a limited amount of capital [*despite being a public company registered in London*] we had to invest in our new facilities from the cash flow from our existing shipbuilding programme, at a disappointingly slow pace. Our reconstruction of Fairfield took over five years when we should have done it in half the time. I think this also applied to other firms.

I know that the record will show that in other countries money was found from outside the individual companies to invest on a huge scale on new facilities. It was found in Sweden, in Germany, in Japan and of course the best example of all in the later years was Korea. Yards like Daewoo and Hyundai were built by the government and handed over to shipbuilders to operate. They did not have to find a penny piece for these great yards.

On foreign competition, much of it was unfair. In Germany, the various Lander have contributed enormous sums to keep the shipbuilding industries going. They do not admit it very readily, but there is no doubt that this money has been, and is, being provided. Much the same thing happened in France, Italy, Spain, and certainly in Korea. In Japan with their own unique ways of doing things, the shipyards certainly got assistance of one kind or another. I believe we were the one major shipbuilding country that played everything with a straight bat-except for the so-called Intervention Fund, which was too little, too late and made available in the wrong form. We have not really had a lot of help from the British Government and certainly in the last ten years we have had no help at all. Thatcher has managed to extinguish the shipbuilding industry, which I am convinced was always her intention. That international competition was the fundamental problem facing the industry was self evident. If you do not get work then you do not survive. What puts you out of business is competition, but there is nothing wrong with that if all players are on a level playing field, we were not.

On labour, in the early days shipbuilding was a rough, rough job. It was dangerous and the environmental conditions were awful. It bred people who were rough and tough. They were not overly well paid, and because of the multi-union structure they were able to use the leapfrogging technique so that any given time there was at least one group of workers trying to catch up with another group. It was a natural hotbed of insurrection and argument. The only thing that eventually slowed the pace of industrial troubles, and eventually brought them to a halt was the gradual realisation that work was running out if things continued unchanged everybody would be out of a job. In the 1950s and 1960s when a big ship was finishing, men were paid off, thrown on the stones, in union parlance, hundreds at a time on a Friday night. We did try to have a rough liaison with other yards on the Upper Clyde to try to dovetail surplus men in one yard with shortages in another. But the lads were not treated in a very humane manner. Instant dismissal meant just that. It was difficult the way we were managing the shipyards then, without any particular attention to product planning, to see how we could have done things differently. The last thing we thought about was trying to plan the order book to provide an even flow of work for the various stages of production. The modern Fairfield and Harland's have a steel working group who are flexible and interchangeable in their approach to work. It allows for an intrinsically more stable workforce because you can assign people to this job one week and that job in another week. Formerly, this would have been quite impossible. We have not got to the stage of having a generic class of outfitters, which is rather more difficult.

At A&P we built SD14 standard ships and B26 and occasionally B35 bulk carriers. Anybody could have anything they liked so long as it one of these three types. When we were building nothing but the SD14 at our Bartrams yard and at Southwick, it was very easy. We launched one a month. Sometimes we were able to squeeze thirteen into the year. You could accurately predict the state of the ship under construction by looking at the date. That sort of production planned itself really. At Harland's we rebuilt the yard to cater for large tankers and bulk carriers. When I first came to the yard I had a look around with Dr. Denis Rebbeck, and told him that the yard was an absolute natural for development. All that was needed was to brick the Musgrave Channel with concrete, put a gate across the end, and we would have the world's largest building dock, and, in effect, that is what we eventually did [*with the aid of the Shipbuilding Industry Board*]. The oil crisis came out of the blue, and virtually destroyed our plans overnight. From then on we were struggling to get any order we could.

At Fairfield we did not really have any product policy. We went to and fro to all our old customers, but when these orders dried up, we would take any order we could for any type of ship we were capable of building. Since we could build almost anything, naval or merchant, we had a very wide field of choice, but it was disastrously uneconomical. I have a picture showing the Fairfield outfitting basin; the passenger liner, *Empress of Britain*, nearly completed, HMS *Blake*, a cruiser, and a frigate, HMS *Chichester*. Now there is no way you can operate a shipbuilding facility efficiently if you are putting together a heterogeneous bunch of stuff like that.

I regarded the Fairfield Experiment as a total unmitigated disaster; a huge piece of industrial deceit. Iain Stewart, whom I know very well, had not been very successful running his own company, Hall Thermotank. He brought in someone from Singers [*sewing machine manufacturers*] as a management expert [*Jim Houston*]. This "expert" produced a string of figures to show how productivity was going up. He was handing out money as if there was no tomorrow. These figures were all rubbish. They were losing money at twice the rate at before the so-called Experiment. It was a total con trick. Whether Iain Stewart appreciated this I do not know. The man who came in from Singers must have known exactly what he was doing. The Fairfield Experiment inflicted far more damage to a company which was already sufficiently damaged. Before Tony Hepper took the job at UCS I advised him to think again, as he would probably get his fingers and his toes burnt. When the whole situation blew up in his face a year or two later I forsook the temptation to say I told you so. He was an example of a number of people who had achieved eminence in other fields, put in by government to save shipbuilding from itself. They all fell flat on their faces.

A&P ran itself, but eventually this strength became a weakness. After fifteen years of building the SD14 and B26 we had no design resources adequate to enable us to prepare for a market once the ultra successful SD14 had run its course. During those fifteen years we made a lot of money when others with more variegated programmes were losing out.

If other British yards had concentrated on standard ships, not necessarily a single type, but no more that two or three, they would have reduced their costs, and probably have lasted longer. Some would, I believe, have survived. But for others it might merely have been a life extension programme. In particular, if the Government had not been determined to extinguish British merchant shipbuilding, and Henning Oldendorff had been allowed to buy A&P, Sunderland would still be building ships today.

I think that we as an industry got more Government support in the earlier days before nationalisation. To a great extent, it depended on something on the face of it that might appear rather trivial, which was actually rather important. That was the character of the person running the particular department in first of all the Ministry of Technology, then the Department of Industry, and then the Department of Trade and Industry. We had one or two who could see our point of view, and could understand our problems, and took actions that were quite helpful. But these guys were eventually sussed out by their colleagues, and they were smartly moved upwards, sideways or out. Civil Servants who had some sympathy with the industry were unpopular with their colleagues, at least that is how it seemed to me. Perhaps that is too cynical a view, but I have never had much time for civil servants or ministers.

After nationalisation, we were all interfered with to a greater or lesser extent by people in British Shipbuilders Headquarters, most of whom were just a bunch of amateurs, and some of whom hardly knew the difference between stem and stern. I was fairly successful at keeping all this at arms length, because I had been rather successful commercially [*not at Fairfield or Harland and Wolff*] and had a reputation, rightly or wrongly, of knowing something about the industry. Thus, they left me on my own. I could not have tolerated some of the nonsense that some of the other managers were subjected to. But it all gradually swept over us like a wet fog. I do not think that British Shipbuilders did much for the industry apart from providing money for certain yards to hang on for a sight longer than they would have done otherwise.

Beforehand, the Shipbuilding Industry Board was a complete waste of time [*despite giving Harland and Wolff millions of pounds*]. They had that man who came from Vauxhall Cars [*Sir William Swallow, a former Chairman of Vauxhall Motors*] who was pretty inept. He came knowing nothing about ships, and never showed any sign of knowing anything about ships. Earlier, Geddes did not do a bad job despite the make-up of his committee [*there were no shipbuilders on it*]. I know that everyone thinks that their industry is different from all other. Nevertheless, having heard this a dozen times as an ex-consultant, I believe the shipbuilding industry is sufficiently different to make it very difficult for someone coming in from the outside in just a few months to get to grips with it, and make fundamental recommendations for change. Some of the recommendations were sensible enough, but others were wide of the mark. Groupings were tried later under [*and before*] nationalisation and were rather a disaster. There was an attempt, which I resisted as long as I was there, to amalgamate the yards on the Wear. Their cultures were entirely different. I knew the trouble that we had in merging A&P and Bartrams. When we did the development at A&P we closed down Bartrams, which had a fine spirit but which was falling to pieces. These two yards were already under a single board of directors, our own kith and kin. But the personnel problems we had were astonishing. I do not think that you can throw yards together, and treat their personnel, blue and white collar workers, as if they were alike as peas in a pod. The older the yards, the more their individual traditions and cultures have become ingrained. I think that the industry was investigated eight times-I think it was eight-and lived to tell the tale. Nationalisation was not necessary for A&P, it was much more necessary, for say, Lithgows; whose financial state was pretty horrific. I think the problem was that the Treasury could not swallow the idea indefinitely that public money could be safely handled by people outside their control, that is, private industry. Hence nationalisation, which did prolong the life of the industry a bit, but when I say a bit, that is exactly what I mean. A lot of things that they were trying to do like central marketing were pretty disastrous. All these people in Headquarters were thinking how do I get myself an assistant? How do I build my own little empire? How do I move up this hierarchy? They started interfering on the most massive scale. I was well aware that my senior staff spent endless hours filling out forms for these guys in Benton House [*Newcastle*] and Knightsbridge [*London*].

This is not the complete answer but I think the most effective thing that could have been done to keep the healthier parts of British shipbuilding alive would have been to disburse the Intervention Fund in a different way. I always believed that applying x percent for every ship regardless was wrong. What I wanted was to have a certain quantum of Intervention Fund at my disposal, which might have been based on our turnover the previous year or on some similar volume index. We should have been allowed to use this sum in whatever way we wanted. If we could have varied it according to our individual circumstances it would have led to a lot more work coming into the industry. It might have enabled us to hang on throughout the shipbuilding recession which is now ending. I argued this at the time, but unavailingly. Nationalisation would have been unnecessary.

Immediately after nationalisation we lost control of our individual workforces. Shipyard shop stewards, by and large, are a pretty shrewd bunch. Over in Belfast they were a really intelligent lot and it was almost a pleasure to sit across the table, compared to the men on the Upper Clyde who were most obstinate. They all knew, in any particular yard, where the end of their road lay. Usually it was with the Shipyard Manager, the chap who was doing all the negotiating. Once we got to nationalisation-there was a whole string of industrial relations amateurs at Benton House-we lost a grip on the whole industrial relations situation. The lads in our yard would now never tackle us for an answer, go back and ask your dad [*at Benton House*] or we will go and ask for you, was the attitude.

Interviews
Barrow-in-Furness

Interview No. 26: Bill Richardson, Managing Director Vickers Armstrong, Barrow and Walker on Tyne, Deputy Managing Director, Swan Hunter, Board Member British Shipbuilders

I was educated at Ocean Road Boys School, followed by Junior Technical and Barrow Technical Colleges, where, during my apprenticeship, I obtained a Higher National Certificate in Naval Architecture with distinction and a Higher National Certificate in Mechanical Engineering, also with distinction. I began my apprenticeship in the 1930s at Vickers as a shipwright in the Mould Loft, before moving to the Ship Drawing Office where I qualified as a Ship Draughtsman. I then spent five years during the war with the Admiralty in London, during which I served overseas in Australia and the Far East. By 1946, I returned to Vickers at Barrow, and was appointed assistant manager in 1952, deputy shipyard manager in 1961, and shipyard manager in 1963. The following year I was appointed Director and General Manager, and relinquished the post for a similar one at Vickers Armstrong's Naval Yard at Newcastle to correct a particularly adverse trend there in relation to profitability and delivery. In 1967, I was seconded by Vickers to become Deputy Chairman of Swan Hunter, during the formative period of grouping on the Tyne, leaving in 1969 to become Managing Director of Vickers Shipbuilding. In 1975 I was appointed Deputy Chairman, and a year later, Chairman. On nationalisation, I was appointed to the Board of British Shipbuilders, and in 1978 I became Chairman of Barclay Curle Ltd., and Chairman of Vosper Thornycroft UK Ltd. From 1979 to 1982, I was Chairman of Vosper Thornycroft Ship Repairers Ltd. On reaching the normal retirement age of 65 in 1981, I was appointed Deputy Chairman of British Shipbuilders and Executive Vice Chairman of British Shipbuilders Warship Division for a period of two years. I was also reappointed Chairman of Vickers, Barclay Curle and Vosper Thornycroft, and additionally, Brooke Marine before retiring in August 1983.

The major strength of Vickers at Barrow was that it was the only company, which had shown anything like the willingness to devote all its resources to the building of nuclear and Polaris submarines. I think the weaknesses of the firm were strong unthinking trade union leaders, and at the workshop level, almost unquestioning obedience of the workforce to union dictates. This, in turn, led to old practices being continued long after they should have been scrapped in the interests of production, plus all the other ills that resulted from union domination. Union pressures are like this; they will put in for a reduction of hours this week, you know that you can not afford a reduction of hours, so you fight it off, but they do not let it drop. The come back again in six months or in one year and raise the same question as part of a general list of questions. I liken it to a ratchet. As management, you are operating a company under a ratchet when every time these chaps claim something they always get a part of what they claim for. So it means that too much of management's time is taken up dealing with unions. Until the mid 1960s we never had anyone equipped to deal with industrial relations, but now we have a director, and under director and a full team of senior officials. It is all a dreadful waste of money compared to what we used to do, and compared to our competitors, the Japanese, for example.

Generally too many firms were small-scale, and lack of investment was most noticeable. The industry was not making sufficient money to plough it back. There was, undoubtedly some poor management, poor labour relations and poor productivity, high costs and late deliveries. At Vickers we never really caught up with the idea of modern marketing, and during the early 1970s national inflation was an absolute killer. It extended our costs, and destroyed our working records, which were all based on low Bank rates and low inflation.

Trade union protective thinking should have been abolished. The real trouble was that there was no one in the trade unions to equate to good management. Good management lives in the future. There was no one thinking that way in the trade unions, so we were always fighting yesterday's battles. We always tried to get the unions to improve productivity, but talking was of no use. It takes almost a person like Hitler. He made sure that there were no trade divisions, they were all made metal workers, and were to share the jobs. During the war and afterwards, they have always carried on as a total workforce.

Here, the average company could not afford a confrontation with the unions. They could not afford a strike which lasted three months, six months or a year.

At Vickers [*Barrow*] we changed the product to meet the demand [*from government*] as we stopped building ships. We decided that the amount of work for nuclear submarine building, plus higher overheads forbade us to continue with merchant building. Warship building has always meant a higher concentration of effort. The way that it was hidden was that Ministry of Defence orders were always few and far between compared to the bulk of shipbuilders, and therefore one was able to swamp the naval orders beneath the merchant ship orders [*in effect, throughout the industry, the transfer of naval overheads to merchant shipbuilding within the same yard*]. In Vickers we got the opposite message and knew that when overheads started cruising up that it was quality assurance that counted [*the latter was a direct result of Britain's nuclear submarine programme in which Vickers at Barrow was the lead yard*], and by the mid 1960s we had over 300 people in quality assurance. The Government set the quality standards, and we had to make sure that the quality was up to scratch. What we had was a good customer who always paid what he owed you. Even if you had a strike you would still be paid. I think that the certainty of the contracts did help

Nationalisation was really a Labour Party ploy. The idea behind nationalisation was that one overall body can far better organise than a series of private owners. From that point of view they must have had something, but you have to ask yourself what nationalisation did for shipbuilding? It only prolonged the end. Nationalisation was a cruel blow to those of us who had been brought up in the industry, but I understood what was behind it.

Having also understood the nature of our Government, it is doubtful that they could do anything. I think that the major change nationalisation brought about was the taking away of power from the man running the company and the replacement of the person and the power by a large, expensive, and unnecessary superstructure. British Shipbuilders Headquarters ran to an overall membership of 300 people. All this was to manage about a dozen yards [*Actually, at July 1977, twenty-seven companies comprising sixteen merchant and composite yards, three warship building yards, five marine engine builders and three training companies. Subsequently, British Shipbuilders acquired one more merchant yard and six ship repair companies*]. There was a great deal of interference from the centre, like standardisation of things such as the cost accounting system, the overheads, labour etc., and it was a ceaseless ploy on behalf of British Shipbuilders to find out what, in effect, our business was. This took a tremendous amount of work, and did not save British shipbuilding. One advantage was that British Shipbuilders brought out a redundancy scheme that helped us to get rid of people who were ready to go.

I was on the Board of British Shipbuilders from its inception until I retired, and was the longest serving member. My first impression of the Board was that there were too many trade union leaders on it; we had three of them. They contributed only on trade union and related matters. I think it was a good thing that the shipyards were eventually privatised because the new look in shipbuilding, shorter delivery and highly competitive cost has, by and large, got rid of the old yard. I recall in the late 1970s the Japanese arriving with their portable computer, and us with our handwritten A4 paper estimating sheets. That made us realise how far we had to move.

The decline started with the trade unions. When I can pick a squad of one hundred men and give them a delivery deadline date, and they meet it without any hold ups or stoppages, this has never been the case. What I would like to do is to discover some means of converting the whole of our trade union movement to an employer plus movement. Given that, you could achieve cuts in costs and in building times. The other thing is that all our winning competitors, the Japanese and Koreans, have beaten us on this ploy. Our workforce is still willing, but they work in too tight pockets that lead to trouble. The Japanese and Koreans bought the whole of a twentieth century technology, using the design forces of the British, or the West Germans etc. If you take the position that they were just starting on their industrial revolution, but additionally had the maximum use of twentieth century devices, how could they lose?

Interview No. 27: Roy Turner, Company Naval Architect, Vickers, Barrow

I joined Vickers at Barrow as an apprentice ship draughtsman, and after two years won a Vickers Armstrong Scholarship. I entered Glasgow University in the late autumn of 1944, and graduated in naval architecture in 1947, and returned to Barrow. In 1949, I was appointed as an Assistant Ship Manager, and in 1950, I was put in charge of repairs, and in the following year I was appointed Assistant in the naval architects department. In April 1957, I was appointed as the Naval Architect to the Company, directly responsible to the Managing Director or Chairman for the overall naval architecture design aspects of all the ships built at the company's two shipyards, the Naval Construction Works, as it was called at Barrow, and the Naval Yard at High Walker on Tyne, a position I held virtually until I retired in 1989.

In the early days the company was run by people who, understanding the nature of the shipbuilding business, knew that there were two people who could get the Board the sack through losses and disasters. One was the chief estimator and the other was the naval architect. Both George Holden and Len Redshaw who were my successive bosses insisted on having a direct line to me. This was an arrangement that was very rare in British shipbuilding. In my view, a lack of similar understanding of the nature of the business led to many of the subsequent technical and financial disasters in the industry.

The first thing about Vickers was that it was a member of a group of companies with initially somewhere in the region of 150,000 employees. It had access to expertise, and, above all, to finance, which was denied to any of the individual companies [*in the industry*]. The most significant example of that financial benefit was the £19 million [*this is a grossly inflated figure, which does not accord with the facts, up to January 1947, Vickers did spend £2.25 million*] investment made in the period 1946 to 1950 in the new assembly shop and the four berths of the shipyard.

The next advantage was that Vickers at Barrow and on the Tyne had large drawing offices of very competent people. In Barrow, there was, initially, an integrated management of shipbuilders and engineers. On thing that was significant in getting some orders was that we were the owners of the busiest and most effective ship model tank in Europe at St. Albans. This gave a tremendous advantage in terms of tendering and the quality of design. Right from the start there was a clear policy of concentrating on innovation in products and methods of build. Individuals, like Redshaw had a very clear idea of how to run the business, and could see how to exploit technical developments with quantifiable risks. For example, when we were about to tender for the *Oriana*, [*a passenger liner*] which was the largest aluminium welded structure that had been contemplated, Redshaw made two full, across the ship, hundred feet wide panels as prototype to determine whether production could be organised in an economical manner.

On the side of weakness, there was a peculiar attitude towards personnel. At the face to face level all levels of management behaved well. They were honest and had a genuine appreciation of the skills of the workforce. In negotiation with the workforce, however, most high-level managers were devious, contemptuous and untrustworthy. Their contempt of the workforce, in this context, was often expressed publicly. There was this strange dichotomy. It largely sprang from the ignorance and obstructionism of the union leaders, and their local representatives, many of whom were poorly skilled and badly motivated. Some of the management's attitude arose from an industry-wide failure to accept the need for workers to be treated properly, to be regarded as essential and valued members of a team. At the beginning of the period this dichotomy had not spread to relations with staff, and in particular with draughtsmen who were held in high regard. As a result of an appalling piece of industrial relations most draughtsmen joined the draughtsmen union. Thereafter, relations deteriorated, weakening the company's ability to recruit and retain the best technicians.

Initially, the merchant shipbuilders thought solely in terms of responding to sequential demands, principally from UK owners. Their attitude to anything out of the ordinary was typified in their attitude to [*the Greek ship owner*] Niarchos when he visited the UK in the late 1940s. Niarchos had established himself as a major ship owner by buying second-hand T2-tankers, and also by building one or two new tankers in the United States. He wanted to start building in the UK, because of the lower build costs. He and his team toured most of the UK yards, and eventually arrived in Barrow. By that time his team were thoroughly disillusioned with UK yards, and had obviously had a most unfortunate experience in many of the places that they had visited. The principal problem was that, in contrast to large established ship owners, Niarchos main asset was the possession of charters to carry oil from the Persian Gulf to the UK. Advice from financial experts available to Vickers demonstrated clearly that this type of asset was a very reliable basis for negotiating a build contract where the finance would be provided independent from the owner in the form of a mortgage on the ship. The security to the lender would be the value of the charter. On the basis of financial advice, we went ahead, and accepted contracts for four tankers of 33,000 tons deadweight. Eventually, we built four more at 20,000 tons deadweight, and the then largest tankers in the world, two at 47,000 tons deadweight.

As UK merchant orders disappeared, many builders found themselves inadequate to cope with seeking orders overseas. They had no management structure to handle agents, and foreign negotiations, and all the complexities of taxation systems. UK shipbuilders knew about the changes in the market, but some believed that they would only be temporary. Few were equipped mentally for the total change in marketing that occurred. Under pressures of competition, British ship owners began to look worldwide to exploit the cheapest possible sources of supply. In doing so they, very often, placed British yards who knew them well, under heavy disadvantage. For example, Vickers built five large cargo vessels for UK owners where every detail was to their specification, including the incorporation of structures that were very out of date in our eyes. This led to very heavy costs, and, because the contracts were taken at very keen prices, on the assumption that some modern features would be introduced, to very considerable losses.

Strangely enough, the Japanese, who built two other ships in the same class, lost more money than Vickers did. However, the next order for ships of that type from the same owner went to Japanese yards with standard Japanese designs. The whole of the owner's preferences and dictated features were thrown aside because of the financial advantage of accepting the builder's standard design. So, in a very short period, certainly less than ten years, the owners swung around from total domination to subservience to what was available from the market. Reacting to that change in fundamental relationships was a great shock to many British yards who could not adapt in time because they did not have sufficient detailed standard designs to compete. On the whole, UK owners demanded higher standards of specification and quality in the UK than they accepted from the Japanese. Standard designs were all important to the Japanese industry in cutting costs and making profits.

In 1966, the Fairfield Experiment was starting to prove the benefits of management and unions working together. The Experiment was, in my view, deliberately stopped before this became irrefutable, because UK shipbuilders wanted to retain their bullying tactics, and the union leaders who opposed devolution of power to the workplace supported the quick shutdown of the Experiment. We had to wait for Nissan and Honda [*Japanese car manufacturers in the UK*] to show how right Iain Stewart had been twenty years too soon. We knew quite a bit about Stewart's work at Govan. Had it been allowed to go on for another six months it would have shown the advantages. It would have been an example, not just for shipbuilding, but for UK industry generally. We would have outflanked the Japanese before they really started.

Restrictive practices were appalling in ship repair, but largely innocuous in shipbuilding, providing there was continuity of production. In other words, if there was a succession of ships to build you could plan your labour allocation so that you took full advantage of the overlapping of the various tasks on the various ships. To some extent the different skills needed were incompatible, and therefore, demarcation was essential. There were too many trades, a feature stemming from the personal ambitions of the union officials. Very many of the national union officials were quite intelligent and co-operative people. It was not in the interests of the local, politically motivated, shop stewards to lose any of their power, which they would do if they had to work together. Some Ship Managers were more successful than others in installing a team spirit. Other managers seemed to enjoy, in the early days, exploiting the differences between the trades, and welcomed disputes as a means of covering up their own deficiencies in planning and in monitoring progress.

On the quality of the management, Len Redshaw was regarded across Europe as one of the best shipbuilders. Many of the other managers were just bullies with little technical knowledge that relied, almost entirely, on force of character. Few managers had any business training, and thus fell victim to all sorts of bad advice from consultants. A manager of the Naval Yard at High Walker was a classic example of the latter failing. Despite being a highly intelligent and a charming personality to his friends, he isolated himself from what was going on in the yard, and eventually pressure from the owners and from the London office forced his resignation. When he was eventually replaced, and managers were sent over to the Tyne from Barrow, they discovered extraordinary things like there had not been a manager on some of the ships for six months. There was no relation between the planning documents, ostensibly showing progress, and the reality outside in the yard. This was largely due to the emphasis placed by consultants on the theoretical approach to a management of what was an essentially practical industry.

Redshaw was very keen, even to a very late stage, to retain a merchant shipbuilding capability. We had played a large part in developing new technologies including the *Methane Princess*, the world's first commercial sea-going liquid methane gas carrier. However, his longer term view was dominated by the need to keep a close relationship with the Ministry of Defence, particularly once the nuclear submarine programme started. For example, at some stage during the early Polaris building programme, we had an order from British Petroleum for an offshore oil platform, which we turned down because the Polaris programme was accelerating. BP was one of our best ship owner customers, and we never built for them again. We never had any opportunity to get back into the oil platform market, but the relationship between Vickers and the Ministry of Defence was preserved.

Was international competition unfair, I do not know what you mean by fair. The Japanese government took over fifty percent of world shipbuilding capacity and in doing so decided to fund vast new yards on credit on a thirty year basis with two percent interest. They directed the brightest graduates into the industry, and they dissuaded active trade unionism, and supported international sales by expert embassy services. This is only sensible government policy if you want to succeed. Whether it is fair or not depends on your viewpoint. UK policy and death by a thousand cuts was totally unproductive. If a government wants to support an industry which is suffering from foreign competition largely supported by taxpayers in another country, then it can not do that by standing aside. The British Government may not have come to the decision that they did not want a shipbuilding industry, but all their actions, though, resulted in the industry largely disappearing. Japanese owners never ordered abroad for fear of disgracing their nation. There were no laws, but there was compulsion by moral persuasion. This is how I summarise the situation: a large nation with a disciplined and intelligent labour force decided to take over a large proportion of the world's shipbuilding capacity, and did so.

I think the UK shipbuilding industry could have expected the British Government to have had, like all its predecessors from Alfred the Great onwards, a policy towards ships and the sea. Such a policy was almost entirely lacking post World War Two, and certainly absent altogether from the 1970s when shipbuilding was characterised as a sunset industry. In consequence, whereas our competitors were being assisted in every conceivable way to increase their share of the market, the British Government relied purely on competitive balance, and tried to reduce their involvement to a minimum. The initial ideas of nationalisation were an attempt to save the industry from extinction. For various reasons they were unsuccessful. Indeed, the influence of British governments on shipbuilding has been deplorable. Neither Labour nor Conservative governments had a policy in regard to ships and the sea. Labour policy was purely based on keeping down unemployment. It was an attempt to secure employment without an understanding of the overall business that was needed. It was dealing with palliatives rather than the fundamental problems.

Nationalisation was desirable in the sense that it was a means by which we could combine our strengths and our financial resources, and present a unified face to government and influence events. The way it turned out, nationalisation was a disaster for the industry. This was partly because of political decisions, particularly by the Conservative Government regarding nationalised industries in general. It was also because, generally speaking, the management at the top of British Shipbuilders was irrelevant to the type of problems which they ought to have been tackling. Additionally, they were, in the early stages, very heavy expenses incurred in setting up a grossly overmanned headquarters, which led to widespread cynicism at all levels in the yards regarding the way in which money was being wasted. There certainly was waste at headquarters. I do not believe that there was too much reporting, but I believe that there was far too much nugatory activity, organising things centrally that could well have been left to groups of yards working together. The biggest change that nationalisation brought to Vickers [*by this stage solely concentrated at Barrow*] was obviously that we were attached to a branch of government instead of a commercially motivated, independent company. Hence the freedom to operate was dramatically reduced. The worst change was that we were forced to come out of the commercial market altogether. We were then forced to come out of the surface warship market, and to concentrate solely on submarines. This last constraint could have the most serious consequences for the future of the company. It means that we are now a single product, single customer [*Ministry of Defence*] company. There are very few examples in history of such a company lasting a long time. With the Government intent on only using one yard for the construction of Trident submarines then the resources required to meet that programme automatically meant that there was virtually nothing available for any other projects.

Interviews
The South Coast

Interview No. 28: Peter Usher, Technical Director, Managing Director, Chief Executive, and Chairman, Vosper Thornycroft

I began my career at Chatham Dockyard as a shipwright apprentice. I graduated to become a member of the Royal Corps of Naval Constructors, and had five years of training. I then joined the Ministry of Defence at Bath as an assistant constructor in 1951. In 1966 I saw an advertisement for a technical director at Vosper Thornycroft, was successful in my application, and subsequently held that post for eight years. I then moved into production for eight years and took over as Managing Director working under a Chairman appointed by British Shipbuilders from 1981 until the firm was privatised in 1985. I led the management buyout bid for Vosper Thornycroft, and in 1985 we succeeded and bought the company. I was the Chief Executive when we floated the company on the Stock Exchange, and became Chairman in 1990, a post I still hold today.

Vosper Thornycroft was created by a merger in 1966 of Vosper Limited, a Portsmouth-based company, and Thornycroft, a Southampton based company. I joined them as the two companies merged. The major strength that we had then and still have is a foothold into the export market for warships, which was very new in those days. We achieved good orders throughout the period of the late 1960s and early 1970s. We were growing at something like a rate of twenty percent per annum, and when nationalisation came in 1977, not surprisingly it was strongly resisted by the major shareholders led by Sir David Brown. Our case was that we had a good order book, a good cash position, and we could see, therefore, no advantage in being nationalised. No one at the company felt that we had a weakness, but I suppose relative size was one. We were making about £3m to £4m profit. We were not a huge company, but we felt that we were big enough to survive on our own. It could be said, however, that our weakness was that we were overtrading. We had a very large order book, and were insufficiently capitalised. Vosper Thornycroft was different because we were orientated to the export market selling small fast patrol boats etc. In the late 1960s we invested, and in 1970 put up a covered berth, which is one of the best in the country today.

Relative to the Japanese, I think we were all using obsolete technology. Generally, there was poor management in the industry, but management spent too much time in dealing with labour matters. There was no time to talk about the product and how best to make it. Every minute in the day in production management was spent on dealing with problems caused by the unions. We had good labour relations compared to those in the North, but anyone could have good labour relations if they kept giving in to the unions. What everyone had in those days, relative to our foreign competitors, was poor productivity. I went to Bath Ironworks in Maine around 1978 and saw productivity there that I was able to measure as something like eighty percent better than our own. I came back and said so, and nobody believed it! Later on when we were getting information out of Japan, where, clearly, they had good productivity, slowly these figures were accepted. The success story that has happened over the time of my career is that we now have overcome that legacy, and we now have doubled or more than doubled our productivity.

I think we achieved that essentially by communication. There was a long period of convincing people that the position was as bad as I recognised it to be, and second, the general change in attitudes. It was also about getting the trade unions off our backs. The trade unions did nothing for the workpeople by encouraging restrictive practices. People in our yard do not run about, but just work consistently. This is what everyone who went abroad came back saying that they had seen in America and in Japan. In our favour, we had a sales department orientated to obtaining export work for years. So far as we were concerned the strongest competition was from European shipbuilders, particularly the Germans. There was, and is, aggressive selling by our competitors.

If you were in a British shipyard twenty years ago, you would see a mass of people wandering about, either with pipes over their shoulders, or possibly not even that. If you challenged them as to what they were doing, they would answer that were looking for components. Shipbuilding is largely a process of assembly of components. A computer-based materials management system whereby you know what parts you need and where they are was not available. A lot of the workers effort went into finding components. The boy or the mate would be sent off to find the piece or wood or a bolt. Productivity is not just about chasing people, although it has to be said that the working day was appalling, starting at 7:30 a.m. and not doing anything until 8:15 a.m., and then slowing down for a 9:30 a.m. tea break about 9 o'clock. There were three components, not working the full working day; not having the pieces to do the work, and there were in many cases old fashioned pieces of equipment that people had to work with. If they were doing steelwork, then it is quite likely that the machine that bent the steel to shape was quite inaccurate. There was many a flogging hammer put on a job to make it fit [*hence the marine engineers' description of shipbuilders as metal bashers*]. Today we would have a cutter that cuts precisely. So these are the three main areas of difference; investment in the equipment, investment in materials management, and working the whole day.

There is something else I should have mentioned; having far too many men each doing only part of the job. It is impossible to organise a burner and a welder, and a caulker as well as all the joiners and drillers etc., all to be there to do their little bit just at the right time. Restrictive practices were at the heart of all our problems, they were appalling, and it is impossible to exaggerate the problem. By the time we had Bill Richardson with us, he was amazed that we were addressing our workforce, face to face, twice a year. In the end we achieved that by sheer persistence as the shop stewards' resistance to it was so strong. It was very difficult to get the workpeople to sit in a room and listen to us for half an hour. We dealt with the problem of restrictive practices by arguing that if a person was able to a job then he should be allowed to do so. The argument for that, simply, was that it made the job more interesting. The trade union member should be there to get an adequate reward and job interest. What they would not accept in the early days was that job interest was something that they had an interest in. I felt that the issue of job interest was our best line of attack. Then men would frequently say that they could do another part of that job. Ok there is no shop steward around, so I will do it. The stewards were the lynchpin of restrictive practices, and the reason why they continued for so long. There was nothing rational in the conduct of the workforce, and many men knew that hotheads were leading them in the wrong direction. The argument that the trade unions used was that if today there are four of us standing around then four of us are employed. That was fine for the job in hand, but it meant that we would not win the next job because our price was too high. I believe that the fact that we were an exporter enabled that message to get across more clearly. Most people felt that work for the Ministry of Defence was essentially given to you. It was your own government, and we did not really compete for it. They could see that we competed for export work.

We certainly had a situation for many years where the full time officials and the national officials of trade unions would denigrate the shop stewards to us by saying that the stewards had gone over the top and were unreasonable. They found, I think, that there power base was being taken over by the shop stewards. They were not, however, man enough on the whole to tell the shop stewards to back off. Full time officials were not prepared to overrule shop stewards.

On management training we gradually moved by getting our foremen to do the National Examination Board for Supervisory Studies courses. We went from virtually no training to a position where all our managers were trained in some way. We spent 0.4 percent of our turnover on training. It may not be enough. I think it was about a half of German levels. In a typical German industry about five percent of their employees at any one time are being trained. In our company it was approximately 2.5 percent.

On senior management, from my experience, I loved old John Rannie at Clydebank, but he was more of a middle manager than a managing director.

Eric Yarrow was much more a managing director in that he was autocratic, but a well qualified senior man. His role, as he saw it, was to get work out of the Ministry of Defence, and he performed it very well. On the whole he was knowledgeable and certainly not an absentee manager. Neither was John Rix, who was the equivalent Managing Director of Vosper Thornycroft. These men worked hard. Rix was extremely strong on customer relations, on selling. The fact that they might have not pressed training, seeing it as an overhead they could well do without, was a product of the times. Training was not a buzz-word thirty years ago.

I have to say that I think our European competitors subsidised their yards and probably still do so. There have been occasions in our experience when a quote from Germany appeared only to cover the materials we had to buy. We, therefore, were convinced that there were subsidies in their prices. The Americans, incidentally have legislation which prevents us or anyone else building for their navy, that is a good piece of protectionism for a start. France has, on the whole, nationalised shipbuilding, you simply do not know what their costs are, but on subsidies, most of them are hidden and the evidence remains largely circumstantial.

I came away from nationalisation after eight years absolutely convinced it was not the answer. It was the whole hierarchy of civil servants and ministers. In the end, issues ended up in Cabinet when frankly they would never have reached managing director level. On the whole, British Shipbuilders tended to call large meetings where the intention was to spread the work around. That takes away the incentive. Why are we doing all this marketing only for someone else to get some work? It de-motivated our sales people, and I think it also de-motivated the whole of our company, because we did not believe that we were getting anything that we could not have otherwise have got.

I think British Shipbuilders had a central objective which they were not prepared to say too much about and that was to run the industry down in size. We started with 88,000 people in 1977. They seemed more interested in reducing numbers than in winning orders, but would not grasp the nettle of redundancy. We ourselves were employing about 5500 workers combined in shipbuilding and ship repair. One of the things that they did straight away was to separate shipbuilding from repair. That did us no good at all. It meant that we had two different organisations reporting to London or Newcastle instead of one. The heart of the problem was that they did not improve productivity. Central pay bargaining was disastrous, and took away the power of local company management. On privatisation, we were able to reduce our workforce. We privatised the company with 3000 people and unfortunately had to make 1000 redundant. We took two thirds of our workforce and ran the place as before, which created massive savings, and gave us cash we otherwise would not have had. We have motivated everyone to believe in themselves.

This would not have happened in this country if Margaret Thatcher had not beaten [*Arthur*] Scargill [*leader of the National Union of Mineworkers*] and the miners strike. There is no doubt in my mind that we have had a civil war in this country. Because Maggie won, managers have got back to managing sensibly. They now spend a lot of time helping their workpeople, not lashing them.

Interviews
The Humber

Interview No. 29: John Leathard, Managing Director, Richard Dunston, Hessle, Humberside, and Director, A&P Appledore Consultants

I started as an apprentice in the Neptune Yard of Swan Hunter in 1944, and began a course of study in naval architecture, graduating with an honours degree in 1947, and then got the 1851 Scholarship from the Institute of Naval Architects to do a research course for a further two years, which I did mostly at the National Physical Laboratory, Ship Division. On finishing there, I went back into the industry for a while with Thornycroft, but then left to work with the Northern Aluminium Company for three years, before moving to consulting naval architecture work with Bernard Collett and Partners. I stayed there for six years, before going to Richard Dunston on the Humber in 1961 as their Naval Architect. I progressed to Technical Director and General Manager to Managing Director leaving them in 1971 to join A&P Appledore International as a Project Director to start with, before becoming a Director.

Richard Dunston was a company, which had to build its reputation on a fairly restricted numbers of types of ships. For instance, I think it was the leading tug builder in the country. It was also quite a large builder of fishing vessels, and other work vessels like dredgers and small coasters. I think it had fallen behind, to some extent, on its design approach. I noticed that when I got there in 1961, that I felt that there were a lot of things that I could do to improve their designs. We tried to introduce them and after a while had an effective design office.

If we look at one of the failures of the company, and of the industry as a whole, the relationship between design and production was not looked at as effectively as it should have been. Design tended to be an end within itself. You did look at simple things, but this was done without any specific detailed approach, you just did it as an *ad hoc* arrangement. Nowadays there is a much greater understanding of the relationship between the two. If you start from scratch you should never put your pencil to paper before thinking of the effect of that on the design of the ship, but also on the construction of the ship. The two things must be considered together. There was no specific attention given to this in the early days, and this had quite a far reaching influence as it affected, to some extent, the way the ship was built. You could have simplified the type of design, and you could have simplified the way the ship was put together, therefore reducing the costs and resources needed to build the ship.

This was true of the industry generally. Now design and production are closely interlinked.

Investment at Dunston was very much limited. Also it was a private company, and one tended to get somewhat restricted views as to how the company should be developed, and as to what the investment should be within the company. We did invest. There were certain things done in the shipyard, but I am not sure that they were done within the consistent approach that they ought to have been. For example, one might suggest putting in a new frame bender, for instance. If you looked at the frame bender in isolation, it could cut the cost of turning a frame by fifty percent, but I do not think anyone had got the point of asking what influence that had on the overall performance of the shipyard. If you increase the throughput of frames by fifty percent, this might totally unbalance the rest of the output of the steel shops. There was no detailed investigation carried out as to what the overall effect was of doing something. I was as guilty of that as anyone else. Here, we are talking specifically about a method of production.

At Dunston's we had a Sales Director, and he and I used to deal with sales and marketing between us. But, it was very largely negative selling, that is, you waited until enquiries came in to you and then you dealt with them. There was a certain amount of positive selling, but not a great deal, primarily because the shipyard had its own customers who kept coming back to it. It had a very good circle of clients, and even in the bad times there were still orders coming in from people that you knew. I think, that to a very large extent, a yard like Dunston's, a small yard, there was not the same need to try and go out and sell. The capacity limitation was somewhere between 2000 to 2500 tons of steel per year, which was, at that time, six large tugs. We could produce up to that steel level. If we could produce six hulls, then we could certainly outfit six hulls, so there was no problem of imbalance within the yard, and we were going through a period where the design and size of tugs was increasing quite rapidly, but the yard was perfectly capable of handling them.

Dunston's did very well; it was quite competitive in relation to other yards in the UK who were building similar types of vessels. We had a very experienced labour force, and it also had an unusual type of method of payment in that there were contracts awarded for parts of the hull to be completed; framing contracts, deck contracts etc. Hours of work content would be agreed, and if people bettered these hours of work, then they would get a bonus payment. It was quite successful, and did produce a productive organisation. At that moment we felt that there should be different types of organisational structure in a yard, that we should be changing from the traditional type. If you change the methods of production from the sequential to the method of building outfit and steel at the same time, then you need an organisational structure that can handle that. This means that instead of having a steel manager and an outfit manager, you may want a manufacturing manager or an assembly manager. This means that you must have multi trade supervision and management of some sort. We tried this at Dunston's in a very simple way. It was very difficult to get people to accept and understand this. Dunston's was a family yard, which meant that even the Managing Director operated within the confined limits set by the traditional owners of the company.

I always thought that labour relations at Dunston's were quite good. We were dealing with; mind you, a small yard. The total number of personnel was about 330, of which 260 to 270 were direct personnel, and the rest indirect. We did have the odd strike, but I always used to feel that the sort of work which people are doing, particularly the very repetitive ones like welding, inevitably call for a bit of an explosion at some time or another. People get fed up with what they are doing. I remember a welder saying that he had to spend his day looking at a little bit of light, and I think to expect people to go on day after day without taking some kind of explosive action, is expecting a lot. We had one very bad strike, but on the whole, they would last only a few days, and we would be able to sort them out, and people would get going again. We had strong restrictive practices in the yard, but we did try to negotiate them out on a number of occasions. We had interchangeability, and we did get them to some extent, and people were prepared to work along them, but they certainly did not go all the way that we wanted them to. But again, that is understandable because people have got a particular skill and that is what they had to sell on the market, and they were obviously reluctant to let that skill be diluted in any way. Unfortunately, I think it got too far in that they got to the stage where it was impossible to get any sort of flexibility or relaxation on this at all. This, of course, had an effect on the fact that we were trying to get multi trade supervision.

The decline of the industry was a question, fundamentally, of capital resources. Capital resources goes right back to the ship owning fraternity through not supporting the shipbuilding industry to the extent that it should be, and expecting that it is always going to be there to produce ships at ridiculously low prices. Then you get to the next stage that the industry gets to the point where it can not develop itself. It has not got the resources to do it, and you do not get a really strong government strategy to try to correct the situation, which the industry has got in to. It could have been done when it was nationalised, because then there was the possibility that there could have been a clear cut strategy, which could have made the industry effective. Maybe it would have to have been a smaller industry, but it could have flourished. Instead of that, I think, to some extent, what has been implemented has been a piecemeal solution, which has not succeeded.

Interviews
Belfast

Interview No. 30: Dr. Denis Rebbeck, Managing Director, Harland and Wolff

My father [Sir Frederick Rebbeck, Chairman and Managing Director] was in Harland and Wolff from about 1905 to 1962 when he retired. He was anxious that I went into the company because he said it was a great industry and a great firm. I went up to Cambridge and did the Mechanical Sciences Tripos so I came in the engineering side and not the shipbuilding side of Harland and Wolff, and then served an ordinary apprenticeship. I went on to the Board in 1946, and was the only Director of twelve who had any academic qualifications, and in 1962 I became Managing Director.

If we start the story for me around the early 1930s it is worth saying that the yard was down to a mere 1200 men, and it was nearly falling off the precipice. The other shipyard in Belfast [*Workman Clark*] which had also headed the output of the world on a number of occasions had gone bust in 1928, was revived, and went on again until 1935 before going bust for the last time. When the Admiralty placed a cruiser contract with us in 1934, we started up again. Further orders came in, and as war requirements developed, the yard did extremely well. In 1941, the Luftwaffe came up from Northern France and flattened us. We were sixty percent wiped out. That was a terrifying blow, there were a thousand people killed in Belfast. The yard had to rebuild and get going again as best as it could. We lost a wonderful opportunity to put in bigger cranes etc, as the Admiralty ordered us to put everything back as it was because that was the quickest way back into production. We put back in the cranes that had been installed in 1913. Our total war output was ten percent of the total UK output, so we were important. We were also important to the city of Belfast as we employed 30,000 men. We really established ourselves as the largest shipbuilders in the world. Post-war, the work requirements were enormous due to all the sinking of ships. At one stage, we had four and a half years of orders on our books. The unions thought they were on a pig's back for ever and all we had to do was to go on building ships. The competition was not severe, the prices were pretty good, and the yard ploughed on quite happily.

At the time nearly all of the steel was being diverted to the automobile industry, which was the number one priority for the Government. We made all sorts of pleas but got no support at all. We also had an impossible time with the unions. In those days, we had something like twenty-three different unions. We pointed out that the Swedes only had two and that other countries had a very small number. So here were we with all this damned demarcation, which was an awful burden. Trades were waiting on other trades to drill holes, and do simple things that they could have done themselves. By 1960, I was deeply involved in the Shipbuilding Employers Federation [*the national body responsible for labour matters, established in 1899*], and each of us had to take the turn of being the President. It was no honour, I can assure you! We had spoken to the government about the nonsense of demarcation, and all the different unions, and we had worked for quite a long time on a new set up for the industry in that we would only have three unions. There would be the steel workers, the finishing trades, and the unskilled men. We got nowhere with it at all in our discussions with the trade unions. I can remember Ted Hill, the head of the Boilermakers and head of the Confederation of Shipbuilding and Engineering Unions. Later he went on to the House of Lords. He told me that I was obsessed with demarcation, and that I was going to destroy the industry because I was going to destroy the skilled men. My reply was to inform him that I was not required to do that job as he was going to do it himself.

With twenty-three unions international competition was going to be a joke. I recall we had a strike of platers at that time, which went on for thirteen weeks. They had gone out on strike and we told them that we would not talk to them until they came back. In the end, they got so desperate that Ted Hill rang me up and requested that we meet in London in one of the lesser known parks, and we could sit on a bench, heavily disguised so that no one would know who we were talking to each other, and then we could sort it out. I refused, of course, and eventually the men came back.

We had very good connections with Norway, so much so that we could do the contracts on the back of an envelope, and we kept being invited back by Norwegian owners who were delighted with the quality of our ships. If you were working for someone like Aristotle Onassis, you read the contract twenty-five times with a magnifying glass, but with the Norwegians you did the contract on a piece of paper. In the 1950s the whole international scene was changing because the German shipyards had been revived, the Japanese were being assisted by the Americans, and our trouble was that we could not get the yard modernised, which was so necessary because of pre-fabrication. We did not have the money to modernise, and nobody was willing to lend us the money.

We were building too many different types of ships. For example, we had eighteen building berths. There might be a passenger ship on one, a warship on another, a couple of refrigerated cargo liners and so on. That was wrong. We also built a big drilling rig for BP, which occupied three building berths. We were always too keen to do something different and new. That did not necessarily swell the coffers. It has to be said that our yard had for many years been a passenger shipbuilding yard. It had been laid out for that sort of work with an enormous joiners shop with about 800 people working in it. The specifications were of a very high order. Even as late as 1960, we delivered *Canberra* and three passenger ships for Royal Mail Lines; we were still building these types of ships. That whole market just disappeared, which was an enormous blow to us. We had lost £1.5 million building the *Canberra* [*mainly riveted*]. We also built a large number of warships for the Admiralty. We were in this area against specialists like Yarrow and Vickers, and, to a certain extent, Swan Hunter and Cammell Laird. We were so vast and we were trying to do so much. We should have changed not only the methods of production but also the types of ships that we were building. I was very much aware of this at the time and had tremendous battles with my father about the lack of proper equipment. It was not until he had gone in 1962 that we put in sixty and eighty ton travelling cranes, for example, and later on even larger ones.

We lost out when the market changed and our former bespoke relationships with ship owners came under pressure. We had always been tanker builders when they had been about 12,000 tons deadweight. We realised that the future lay in bigger ships. I did a lot to get the graving dock, as distinct from the building dock, into Belfast because the repair side was always an important part of our facilities. I did lay plans for the building dock. The difficulties were, of course, financing these projects. We had always built fairly ordinary freighters for example, Andrew Weir's Bank Line. They only built with Doxford on the Wear and with us. They never went to Japan or to the continent, and so we treasured that connection. After the war, the traditional customer client relationship remained very strong. We had understandings with various shipbuilders that were you had a client that you could regard as your own they would quote a higher price for orders and keep off your grass as it were. It eventually broke down because a lot of the clients disappeared. For example, we had built every Union Castle Line mail boat for fifty years. Then Clan Line took them over, and decided to go to the market. I remember my father being very upset about this because there had been a very close relationship between him and Sir Vernon Thompson to the extent that they had done contracts on the telephone on a Sunday evening from their homes. Clan Line rang us up and said that they were very sorry, but they had received a price that was so much cheaper than anyone else's that they had to bypass us. It was Cammell Laird who had won the order. They were a firm who had not built a passenger ship since before the war. I heard that they had not done well out of the contract and phoned Robbie Johnson, who was a friend of mine and counterpart at Cammell Laird, to come clean and he told me that the contact was a bloody disaster and that they had lost £1.25 million on it.

Also Port Line, who had been a very good customer; did not require the Empire Food Ships because of the changes in Europe, and the same could be said for Shaw Savill, so particular types of trade were declining. We did talk to the more sensible unions, but some of them would not even sit down at the same table and talk to each other. The shipwrights and the boilermakers were bitter rivals; they hated each others guts. The unions were far too powerful; management was just punch drunk by that stage. We did have wonderful foremen, particularly in the Engine Works; they were the real managers in a sense. They had absolute authority as they could hire and fire, but this was eventually changed when the socialist touch of personnel management eventually came to the fore.

I always thought that international competition was unfair. I knew damn well that there were hidden subsidies in some European countries. Our government did not give a damn about shipbuilding. In the forty years I was at Harland's I never found a sympathetic government. It was always another industry which gained the support. At least, that was my impression.

On the other hand, we had tremendously capable technical staff. We had a very close link for main marine diesel engines with Burmeister and Wain of Copenhagen and were the main licensee for their engines in the UK, and had only one sub licensee, Kincaid of Greenock. We built all our diesel engines, and also those for Cammell Laird and Swan Hunter. We benefited from the popularity of Burmeister and Wain engines, and we also had some extremely good workmen, very loyal.

We used to find that Swan Hunter could undercut us by ten percent because they were more streamlined than us. We were expensive when we built only passenger ships for a captive market, all that mattered was quality. We were very good on delivery; we were never late. We did gild the lily a bit. We could never seem to adjust to what the North East Coast did on the tramp ship.

I liked Geddes. I thought he was sensible and far looking. He wanted us to group with Vickers and Cammell Laird. I was all in favour of that. I saw no virtue in isolation. I had private meetings with Vickers and Cammell Laird but I did not get the backing within my own company. The whole thing floundered. We could have done all the naval work at Barrow, and Cammell Laird would have been strong on the ship repair side. We would have been able to do the bigger ships, the tankers here in Belfast. Our dock could have built a mega ton tanker. As far as mergers on the Clyde were concerned, so much of their industry was out of date that merely having amalgamations would not really achieve anything. It has to be said that it was the market that changed, and it did not really matter how much the yards changed in one sense.

I feel that government influence on shipbuilding was negative. I do not think governments should interfere and put people on boards. They should not fiddle with an industry they know nothing about. I do not think that nationalisation was a good thing.

On discrimination against those of the Roman Catholic persuasion; Harland's had always been a Protestant stronghold. That was a fact before I was born. I believe that there were incidents when Catholics were chased into the tide. You can say that all the trades were staunchly Protestant. Where you might have got some Catholic tradesmen was amongst the painters, the furniture polishers, but never amongst the hard trades. Management turned a blind eye to it because it was tradition, and the foremen did the hiring and firing.

Interview No. 31: T. John Parker, Naval Architect, Harland and Wolff, Managing Director, Austin and Pickersgill, Deputy Chief Executive, British Shipbuilders Plc, Managing Director, Harland and Wolff

I came to Harland and Wolff in 1958, and studied naval architecture at the College of Technology, and partly, at Queen's University, Belfast. On qualifying, I came to work full time in Harland's ship design team, and was there for a number of years before I branched out around 1968, into management. I was subsequently involved in the application of computers in production. We were one of the first companies to adopt the Norwegian Autokon system, which involved the numerical control of cutting machines. At that time we were also constructing the giant building dock, and introducing modern shipbuilding methods. The panel line was to follow, and it was a quite an exciting time of radical change in terms of equipment and methods. I then moved out of production management and into technical management, where I became manager of the production drawing offices around 1971. Then I moved into sales and marketing, and was there during the tenure of the Danish Managing Director, Ivor Hoppe [*ex-Managing Director of Odense Shipyard, Denmark*] for about three years, before moving in 1973 to be Managing Director of Austin and Pickersgill. I was there until I got 'hijacked' in 1977 on nationalisation to British Shipbuilders. I was seconded three days a week by A&P, and remained with British Shipbuilders attaining the post of Deputy Chief Executive, until in 1982, I was invited back to Belfast by the Secretary of State for Northern Ireland, Jim Prior, to tackle the very difficult situation at Harland and Wolff, who were running out of work, and have been in Belfast ever since.

I think that the whole industry suffered from the cost plus mentality of the previous forty or fifty years. The management of the industry were not sufficiently internationally minded. They did not see, or take an interest in, what was happening globally in the 1950s and the 1960s. The threat of Japanese domination was not being taken into account in the 1960s, and that was a period that was the decisive turning point. The product and production development processes going on in the industry lacked attention. Britain led the world in hydrodynamics and in the more esoteric aspects of naval architecture, but I believe it was a preoccupation with that aspect, and not with the production or market sciences, which was a real problem.

For example, in this country there was a tremendous resistance from the chairmen of the day to the change from riveting to welding. Sir Frederick

Rebbeck, I understand, here at Harland and Wolff, was not very well disposed to this fundamental change in production technology. There was a sense of market complacency that orders would always be there. All that started to change in the 1960s, and had markedly changed by the 1970s.

Lack of investment was also a problem. The investment was left too late. These things lie at the door of management. There is no doubt that there was money made in the 1950s and into the early 1960s that could, and should, have been applied in a re-investment sense. Poor labour relations varied a lot from company to company. You had over-manning practices and total dominance by individual unions, which had to be broken down. It is only now that we are seeing the benefits of full flexibility. Equally, I am not sure that the rigid demarcation of trades was not the massive problem that many people would make it out to be. We had rigid demarcation at A&P, but we also had a highly productive organisation. We were slim in management terms, but the quality of the management and supervision was good, and we had strong shop stewards who would always deliver what they negotiated with you, so I have never been against negotiating with strong union representatives, unless it turns into a negative anti-management campaign. By and large, I would say that the unions should not be blamed for the failures in the industry. This is simply too convenient. The quality of the management very often got the type of unions that it deserved.

In management terms, some of it was good in parts, and some of it was lousy in other areas. Undoubtedly, some of the wrong people got to the top. I think that there were some family dynasties that were a problem in a sense that they did not create change. The industry was too stagnant; and if you look back, it comes from the lack of market awareness, and from management not really seeing the international pressures that were likely to be on them tomorrow. It was this stagnation and the lack of movers and shakers, which were prevalent in some areas. In other areas you had well qualified people who managed to get to the top through their sheer professionalism, and strength of leadership. Examples of the latter are Jim Venus who got to the top at Appledore, Derek Kimber, my Chairman at A&P, and Ross Belch who, in his own way, had a lot going for him. To some extent he had the Lithgow and Scott family on his shoulders, and it was difficult to make the radical changes that maybe should have been made there, but he did a lot. By this time, the family dynasties were on the decline, and it left the headroom for the professionally trained technocrat to get through. Even letting him through was not enough. It was giving him the freedom to manage, which many of them did not do. Some were given responsibility, but not authority. Some were also not really exposed to the competition that was building up.

Something had to be done to shake the industry. I do not know if creating bigger conglomerates was necessarily the answer. This is a people business, and whilst it looks wonderful on paper to create something larger out of two wholes, it destroys the competitive spirit of the individual company. It would have been better to modernise and rationalise down to a fewer plants, perhaps, and invested into each of the yards, and placed much greater emphasis on marketing, products and production technology developments.

I think there was a period where it was very difficult to discern a government policy other than that high profile one by John Davies [*the Conservative government's Industry Minister*] over UCS. After that, people tended to run away from shipbuilding, and there were not many who wanted to become involved at ministerial level. There was a great shift in where responsibility for shipbuilding lay. If you had a government that was strategically concerned about its shipping, it would naturally be strategically concerned about its shipbuilding. Sometimes shipping was in trade, sometimes it was in transport, and all over the place. There has never been a marine industry minister in this country, and I think, unlike Japan and France, that has been a weakness. A more co-ordinated policy between shipping and shipbuilding would have been more effective.

I have never been impressed by consultants coming along and saying that small is not beautiful. If they see something big, they want to break it up, and if they see something small, they want to make it bigger. My view would be that it is the quality of the management in the individual company that ultimately matters.

The Labour Government did introduce the tax incentive schemes for ship owners to re-invest, and hopefully order in the UK, but the big weakness in that was that other owners could come in and set up here, and then go and order in Japan. The Labour Government, when dealing with a Secretary of State such as Eric Varley [*who succeeded Tony Benn*] appeared to me, as a manager, to have greater practical perception of the industry. It is not a dirty thing to them, as it is, to some extent, with the Conservatives.

Undoubtedly, the question of competition is a difficult one. There has been a lot of subsidisation in every shipbuilding country in the world, our own included. It is a distortion, and the day it is got rid of, the sooner the better. But, it has played a big part down through the years in distorting competitiveness at one stage or another. I think that it is only now that it is being globally tackled. No one in the shipbuilding industry could have made money when the prices were, in effect, less than the cost of materials. That situation pertained for a number of years during the late 1970s, and during a good part of the 1980s.

By the time nationalisation came in 1977, it was clear that the industry was facing two major crises. One was a financial crisis, partly through its own inefficiency, and partly through the price of ships falling, and the second crisis was the market crisis of demand disappearing. Something radical had to be done. If the industry had just been left to its own devices, a lot of companies would have fallen out of the tree. Nationalisation, as it proved over time, allowed a more controlled rundown of the industry.

At the beginning of British Shipbuilders there was no corporate plan, no blueprint for the future, and we really had to start from scratch with a market that was running downhill. This was post the first oil shocks in 1973-1974, so the global prognosis was still pretty awful for the industry. No one knew just what the cumulative financial position of the individual yard was. It took some time to discover that the loss making was at a remarkably high level. The first accounts for the industry show very significant losses. There was a Labour Government in power with obviously an election looming. It was not the time to start on savage cutbacks of the industry. We did pull a corporate plan together fairly swiftly with Michael Haines, who had been taken on to establish financial facts and controls. I also inherited the Polish Ships deal, when I went there. The distribution of that order from strike-bound Swan Hunter to the Clyde and the Tees, brought time for the industry, and whilst it was heavily criticised at the time, the prices that were obtained, were, in hindsight, fairly good.

The last number of years in the nationalisation process did kill off businesses. The strategy had been one of managed survival. How do you get across the "V" Valley of demand falling, which would ultimately rise? People, having kept their nerve for so long during a controlled rundown with plans for the upturn, suddenly lost it in the Kenneth Clarke era. That is my opinion about nationalisation. It has not happened that way in Germany, or Spain. To some extent in France it has happened, but they still have a large shipbuilding industry there. Germany is the classic example. They have slipped, but they have kept a lot of their shipbuilding horsepower-in a nation that is paying twice the wage rates we are. I feel it was very sad that Kenneth Clarke [*Minister of State*] appeared to lose his nerve, or became so impatient. If they had independently asked someone from the outside to analyse what was happening globally in terms of a potential upturn in the business, they would have discovered independent evidence that I believe might have led them to a different decision. The question continually facing us was how you could control your losses to a minimum extent while you got across the market chasm? The government generally jump from rock to rock. I do not think that there is an industrial policy for any industry today in the country.

When the industry got to a certain size, and when you sufficiently crossed the valley to a point when you were seeing the mountain face of recovery in front of you again, then I think the timing was right to consider some strategic change for the industry. Privatisation was as good a way to go as anything. Personally, I would say that shipbuilding does not fit easily into government ownership. Indeed, there are very few industries that do. I have worked in both the public and the private sector, and I can assure you that you spend too much time in the public sector explaining, as senior management, what you are doing, why you are doing it, and why you want to do it. In the end, they are still liable to take your head off if something goes wrong. Whereas, in the private sector you explain to your Board what you want to do, and they either agree or disagree. If you go wrong, you may still lose your head, but that is ok. In Government the interface is not an easy one. Not because Civil Servants are not highly intellectual people, it is because they are, and when dealing with you as an industrialist, it is their intellect that they apply, not industrial experience. You have to satisfy their intellect by explaining things, and if you explain things it takes a lot of your day up.

I would say that British Shipbuilders, as a headquarters organisation got it wrong. I fought against it when I was there. When Robert Atkinson arrived [*as Chairman*] in 1980, I was then acting Chief Executive. He had a paper from me on his desk the first day he arrived, which steered him through some of the key issues during the first months. This paper advised him to hack the overheads of the centre down; there were too many cars, too many people, and too much paper. In fairness, he did this. Again, in fairness, the problem in a nationalised industry is that you need a lot of people at the centre to gather the information to allow you to "explain" to government. It did become top heavy at headquarters, and did not add value in all areas. I would also argue that it did add value in a number of key areas. For example, you were able to centralise legal services, and have much more grip on research and development for the industry as a whole, to have a much more co-ordinated international marketing effort, and also to have some common approach on systems and productivity improvement drives. Those were practical things that were positives, but it was coupled with a lot of bureaucracy and people at the centre trying to run the business, which was not their business to do. Many just did not have the relevant experience to do so.

I would like to have seen a closer link built between the ship owner and the shipbuilder, to the advantage of both. I would have liked to have seen that pushed in an EEC dimension. I think that there was great scope for the renewal of the EEC fleet [*the fleets of the individual member countries combined*] based on the fact that it was getting old and obsolete. You had a great decline taking place not only in Britain but also in Europe. So, if you put all that together, there was a strategic case made for the ship owner benefiting from slack capacity in the yards, for taking that capacity down at a slower rate, de-manning to get efficiency, but not necessarily getting modern facilities. We have closed two of the most modern yards in Europe in Sunderland, and it should never have been allowed to happen. They were efficient yards, which could have been made much more efficient, and it was all very sad. I think it as a lack of vision on the part of government; and symptomatic of the Conservative administration in that they do not really have any great industrial vision. They did not often understand the very cyclical nature of this business, or the structural problems that the oil shock globally gave to the business of shipping and shipbuilding. There was scope for a scrap and build policy, which would have taken a lot of the old ships out, and would have left Britain and the EEC in a strong position, if new, modern, and low fuel consumption ships had been introduced. The money that they have thrown at closing down facilities, and putting people on the scrap heap could, I believe, have been much more effectively deployed.

Interview No. 32: Eric Mackie, Deputy General Manager, Ship Repair, Harland and Wolff, Belfast, Liverpool, Thames, and Southampton, General Manager Ship Production, Belfast. Managing Director, Brown and Hamer, South Africa, Executive Board Member for Ship Repair, British Shipbuilders Plc, and Chairman and Managing Director, Govan Shipbuilders, Managing Director, Kvaerner Govan

I served my time in a textile engineering company, Mackie of Belfast, and accumulated the qualifications necessary for the Institution of Mechanical Engineers. I then went to sea, and ended up as a Second Engineer. To have been Chief would have meant waiting to fill dead men's shoes. I then came to Harland's as an assistant manager in the ship repair company. I was then appointed as Deputy General Manager, and then General Manager in charge of the Southampton facility, where I remained for two and a half years. I then returned to Harland's in Belfast to take charge of their entire ship repair operation on four separate sites. We employed over 20,000 people in ship repair. On the arrival of the Dane, Ivor Hoppe, I took over ship production, but kept the responsibility for ship repair, closing down the three mainland yards in London, Liverpool, and Southampton in order to concentrate our business, with Government support, in Belfast. After three or four years, I went to South Africa and managed a subsidiary of a large mining company for six or seven years, before being headhunted by British Shipbuilders to head their ship repair companies. I was then asked to go to Govan Shipbuilders who were making tremendous losses, and stayed there until the company was privatised and subsequently spent a further three years with the new owners, Kvaerner.

In Harland's there was a great deal of technical strength and practical management, but we were hardly strong on management skills. Management were always at the will of the owners of a shipyard. Senior management would overrule junior management on the threat of strikes in order to get the work out. They would allow money to be paid, or the conditions to be altered to suit the workforce without trying to control it. On the other hand, it was a terrible thing to see men standing outside the gate waiting for two or three weeks work, before being put out on the street again to join the hordes waiting for work. This had happened for generation after generation, and as a result there was that much distrust between management and workers. They believed that they were merely a tool for the job and would be discarded afterwards. They did not have a say in the industry. This took a long time to overcome, and even yet is still working its way out. Previously, in management terms, all you had to do was to tell the workmen to go and do it or they were out of the door. There was an almost universal lack of communication skills, and demarcation was down to defensive job protection, and there were too many people trying to do the same job. Also, there was a lack of investment; there is no doubt about that. The whole thing was to produce on archaic equipment, get it out, and make money on it; there was nothing ploughed back in terms of re-investment. Obsolete technology in production was a major problem, and marketing was undertaken on the Old Boy Network basis, until British ship owners found that they could no longer afford to build ships in Britain, both at the prices and in the time it took to build ships. Although, there was no doubt that Japanese competition was unfair relative to European shipbuilding. All the Japanese-owned ships were built in Japan; therefore, they always had a base load for their industry, which makes you competitive in export terms, because a lot of your overheads are already covered. In effect, Japan was a closed market, we were not.

We were fortunate in Harland's that we got the money [*from the Shipbuilding Industry Board*] to build our giant building dock, but Belfast was an isolated area. Obviously shipbuilding in the UK needed to be rationalised, and this is where Geddes, and subsequently, the SIB came in. I say this, and it sticks in my throat every time I say it, but Govan Shipbuilders should never have survived because of its geographical position. It could build ships only up to 60,000 to 70,000 tons. It had no dry dock. Every launch was an adventure, and it was a credit to the Scots that they could ever build a ship there. When you look at Scott Lithgow, they had spent money on a tremendous dry dock [*the Firth of Clyde Dry Dock, later known as Scott Lithgow Dry Docks Ltd., at Inchgreen, Greenock*] and then closed it down. That was the finest dry dock in the UK. You could have built the facilities down there [*the Lower Clyde*] that would have made you a fabulous yard. You would have probably had to have closed Yarrow, but nevertheless, you would have had a super duper yard. We missed out on opportunities

I think that there was no way forward other than through nationalisation. It was absolutely essential that it took place. There was no way that all the yards should have been kept open as the market could not have supported it. There were yards that did not deserve to be kept open, and there were yards that were closed which did not deserve to be. At the outset of nationalisation, Graham Day said shipbuilding capacity in Britain should be cut down by forty to forty-five percent, which was a hard, hard situation to accept for yards with ships in them. Before Day went, and British Shipbuilders began in earnest in 1977, the industry was living hand to mouth. People wanted their money out.

Ship repairing should never have been nationalised. It was an industry different to shipbuilding. The workforce in ship repair was one that was traditionally used to high wages and high work intensity for short periods of time. The system [*casualisation of the workforce*] worked admirably in ship repair, and produced very cheap prices. People made a killing, but there were not the overhead structures that carried on afterwards. Therefore, it was beneficial to both the ship owner and the repairer. When some of the ship repair firms were nationalised, they wanted the same guarantee of employment and the same type of conditions as the shipbuilders. Consequently, when the orders were not there, the ship repairers still had to maintain their labour force, which would eat away their profits; and put many of them into liquidation.

I was at British Shipbuilders Headquarters for a relatively short time before being asked to go and try and sort out the situation at Govan Shipbuilders, the old Fairfield yard, because of huge losses, the belief was that the only thing that could be done was to close the yard. I met the directors first of all. Directors are easy people to frighten, because they have a lot to lose. If you hammer the table and say that you are not going to put up with all these losses, they just sit there and nod their head. When you go and see the shop stewards, it was entirely different. Govan had the best set of shop stewards I ever encountered. They were knowledgeable and tough. I always remember Sammy Gilmour when we first met. I thought that what was good for the goose was good for the gander, and started to do the same thing as I had done with the directors. Gilmour interrupted, saying, "just a minute, who are you? What are you? Are you a knickers salesman or a Singer sewing machine man?" These were the type of people who used to be directors. After some questioning on my part, I said to Gilmour, "the next time you use that language in front of me, I will get my size 9 boot and put it where it hurts." Gilmour replied, "we have got a shipbuilder here," and from then on we hit it off.

Instead of making a plan for closure, I made a plan for survival. I told the punters [*workmen*] what I was proposing, and went down to London to meet the Government through British Shipbuilders, and I filibustered for two to two and a half hours with them. They were disgusted with me, and it did not enhance my career one little bit. A man from Finance, Maurice Elderfield, said "on your head be it," and we put a plan together. I thought Govan Shipbuilders was too big to survive the way it was with both the Scotstoun Marine [*the old Connell yard*] and the Govan yard. The market place could only support the Govan yard, and I negotiated the closure of Scotstoun. There was drama, but there were no strikes, and I got it done with the agreement of the unions. Consequently, the workforce, through a redundancy scheme, was reduced from 6000 down to 3200.

We then moved forward to finish of the Polish ships, which was the order that almost brought the entire works down. In defence of Govan, however, they did a marvellous job with the Polish order. They delivered thirteen ships in two and a half years. They did not negotiate the price. The ships were the wrong size for them. Nevertheless, they built and delivered them, and those ships had the lowest guaranteed claims of any in British Shipbuilders. They were magnificent ships, and it was not Govan's fault [*the order was originally transferred from Swan Hunter because of a labour dispute*] that they got cut to pieces with the losses.

When I came to Govan, I was very fortunate in one way, inasmuch as I came after the Fairfield Experiment and the UCS Sit-in [*Work-in*]. Taking the sit-in first, here you had Jimmy Airlie, Jimmy Reid, and to a lesser extent, Sammy Gilmour. Reid said to the men at Clydebank, "there will be no late coming, there will be no sitting down, and there will be no bevvyin [*over consumption of alcohol*]." All of a sudden they realised that they had to impose that discipline, and they began to realise how difficult that was. A lot of people said the Fairfield Experiment was not a success. In my view it was a huge success. When I came to Govan these dramas were all over, the men themselves, along with the management at that time had decided that they would work multi-skilled; full flexibility and interchangeability. Now that was without a payment, without a strike and without an argument. They decided that to get the Polish order finished and to protect their jobs they were quite prepared to accept that. I had, therefore, something to build on.

This was one of the things that made me keen to continue to run Govan. I also found that when I talked to the Convenor [*Chief Shop Steward*] and Shop Stewards, and to the whole company at a Bingo Hall in Govan, they turned up half an hour before work, and they listened. I could identify with them, and see what they wanted out of the yard. I told them that we could make Govan survive on one condition; that we map the plan out. I told them that I was going to close Scotstoun Marine and lose almost half the labour force. I told them as I believed it, that this was the way forward. Gilmour put the point that the workforce was putting its head on the block, but that I was not, as I had a job to go back to. That was the time I decided to give up my job at British Shipbuilders as head of ship repair, and concentrate fully at Govan, on British Shipbuilders behalf. British Shipbuilders did the market research and we went out and got new orders, including bulk carriers from China.

I changed the technology in Govan back to passenger shipbuilding, so that I could attack the North Sea replacement ferries market. Hopefully, from there we would then move on to the cruise ship market. Unfortunately we got no further than building North Sea Ferries, where we came right up against unfair competition. We had the technology in place, we were building modular cabins, were cutting man hours off the ship, and so on. We were right in place for the next ship, the Brittany ferry. We had spent a lot of time and money on the design, and we had many a meeting with the owners, and met all their requirements. The Christmas before the order was announced, I was told by the Managing Director of the owners that the order was secure. Towards the end of January, he rang me up and told that the ferry was to be built at Chantiers de l'Atlantique, who had not even quoted for the job! I was told by an insider that the representatives of Chantiers just walked in and said, "we are going to build that ship for you." They were a nationalised yard, and had two ships to build, but one had been taken away from them, so they wanted the ferry to fill the gap in their order book. There was no estimate and no specification, and time testing as we had done. We had spent somewhere in the region of £1.3m on it.

Our Government had a go at the French Government about it, and got short shrift, because they were basically told to get lost. We took it to the EEC and the European Court at Strasbourg, and we won. We had won the battle but we had lost the war. Where is the ferry being built? The French looked after their own, whereas we did not. There is no doubt that it was unfair competition on a national scale from the French, and from the Japanese. At Govan it was the willingness and the identification of management with the workforce that enabled us to achieve. If we had that type of relationship before, just think how far we could have gone. I had the labour force working for me, and I had the whole of Scotland working for me. The Press worked for me and so did the Scottish Office. The punters worked for me, and they worked for their own yard. That was the collective effort that made Govan survive [*that, and huge amounts of capital investment from previous governments and British Shipbuilders*].

In the industry generally, management and unions contributed very heavily to the others troubles. They were only destroying each other.

Interviews
British Shipbuilding Industry Officials

From a plethora of local employers' organizations, a national body to set rates on pay and conditions, health and safety, etc. was formed in 1899. Based in London, the Shipbuilding Employers Federation (SEF) set policy for the industry and convened national conferences with the myriad trade unions involved. Before reaching the national level, however, yard, local, district and regional conferences were held. This led to a highly bureaucratic system of industrial relations where the worship of procedure became dominant and thus limited flexibility. Before the 1965-1966 Shipbuilding Inquiry Committee Report, there were still eighteen extant local employer federations affiliated to the SEF. On the ship repair side, the Dry Dock Owners and Repairers Central Council had been formed in 1910 as the national coordinating body for myriad local employers associations. It was not until 1928, however, that the shipbuilding employers decided that they needed a trade association. Accordingly, the Shipbuilding Conference was born, primarily as a price-protective cartel but also as the industry's main lobbying body to government. To complete the ambit of the industry, the National Association of Marine Engineers was formed in 1939 as an affiliate to the Shipbuilding Conference. These associations continued up to 1967 when, as a result of recommendations contained in the Shipbuilding Inquiry Committee Report, a Shipbuilders and Repairers National Association (SRNA) was formed, bringing them together in one national body. After opposing the nationalization of the British shipbuilding industry from 1974 onwards, the SRNA was wound up when the vast bulk of the industry (Harland and Wolff, Belfast, being the major exception) was brought under public ownership through British Shipbuilders Plc in July 1977. It then reformed and renamed itself as the Shipbuilders and Ship Repairers Association (SSA). The SSA claimed that though all the main shipyards had gone, there remained in operation some fifty small yards producing small ships.

Interview No. 33: R.D. Brown, Deputy Director, Shipbuilding and Repairers National Association, Director of the Association of West European Shipbuilders

I qualified from Aberdeen University in 1937, and then spent a year at a law firm in Glasgow, before deciding there was little prospect of success there. I came down to London in 1938 and joined the British shipbuilding industry's commercial association called the Shipbuilding Conference [est.1928]. The

Chairman at the time was Sir Amos Ayre, who with his brother, Wilfrid had founded the Burntisland shipyard in Fife after the Great War. The second in command was Alexander Belch, whose son was Ross Belch later Managing Director of Scott Lithgow. I eventually became joint secretary of the SRNA, formed by the amalgamation of the Shipbuilding Conference and the Shipbuilding Employers Federation in the mid 1960s. Eventually, when the senior people died, I became Deputy Director in 1970 or 1971. I served in this capacity until the SRNA was disbanded on the advent of nationalisation in 1977 under its Director, Cliff Baylis. I then retired, refusing to go and to join British Shipbuilders, and then became Director of the Association of West European Shipbuilders until 1983.

One must remember that most of the shipbuilding yards in the UK are upriver, and very few of them lie directly on the sea. The glut of orders post-1945 prevented them from modernising their yards. Most of them were cramped for space on sometimes very narrow rivers, and therefore could not go for the very large ships that the new European shipyards and those in the Far East, who started from nothing, could deal with. This was one of the major reasons why the shipbuilding industry in the UK reduced her chances of survival.

The strength of the industry was the yards were all independent of each other, and yet very co-operative in certain directions, for example, on research. They would help each other in times of need, while still being highly competitive, one against the other. There was little or no co-operation on the financial side, only on the technical side for the presentation to the outside world. One of the major difficulties was the attitude of the trade unions in the post-1945 period.

The real difficulty, in my view, was that when the ships demanded by world ship owners rose in size to the 100,000 to 400,000 deadweight ton level it was impossible for the British shipbuilding industry to compete due to the small size of yards, rivers, and estuaries. British shipbuilding yards had to fight all the harder to maintain their ability to fight for the smaller ships. There was a degree of inevitability that the industry would decline. I do not think that the family dynasty thing was a major problem, although it may have led them to be less adventurous. I think that in any event that the market was vanishing. Smaller shipbuilders did attempt to develop standard ships but the market was limited, and was ultimately in favour of the new producers. I think that both central organisations on the labour and commercial side of the industry did an excellent job in providing information. There was, however, no compulsion that we could exert.

I was opposed to nationalisation because I felt that an enormous superstructure would be created, which would in the longer term increase costs. Yet, one could equally say that the industry was declining and that yards would have to close. The whole theory of nationalisation had a detrimental effect on the management and workforce of the shipyards. The first time around we were able to defeat nationalisation because the SRNA pushed the issue of hybridity. I spent hours of the day and night in the House of Commons at the time of the passage of the Bill helping to organise the opposition to it. We would brief Members of Parliament. The SRNA actively opposed nationalisation.

There were far too many people in British Shipbuilders. Graham Day, when he came back in, to my mind, did an excellent job, although I can imagine that a number of shipbuilders thought he was a bit of a hatchet man. The motivation to nationalisation was ideological [*as was the opposition to it*]. If you had sat in on the meetings with Tony Benn, he would ignore all your arguments and representations. Having been beaten the first time around on hybridity, he was determined to ensure the second Bill went through. I think that the Labour Government may have made promises to the trade unions. They were perhaps under an obligation to nationalise the industry. It may be that they thought that nationalisation was the only way forward.

Interview No. 34: Joseph Charles Asher, Shipbuilding Employers Federation, Shipbuilding Conference, Secretary and Administrative Director of the British Shipbuilding Research Association, and later, the British Ship Research Association

I came into the industry when I was seventeen in 1935, and spent two years with the Shipbuilding Employers Federation, whilst contemporaneously studying for an evening degree at the London School of Economics. I found the procedures for dealing with everything by consulting precedent at the SEF not to my liking. I still think they were wrong. Then I had a lucky break and went to the Shipbuilding Conference with Alexander Belch, who became Deputy Chairman. In 1939, I was made personal assistant to the Chairman, Sir Amos Ayre. During the war he was Director of Merchant Shipbuilding and Repairs, and I stayed with him throughout, until in 1944, the industry decided that it should have a central co-operative research facility, so it formed the BSRA, and I was appointed Secretary in December 1944. In 1955, I was appointed Administrative Director, and just after nationalisation I retired at the age of sixty-one in 1978. So, effectively, I was at the head of BSRA for thirty-four years.

One of the first things I learned when going in, in 1935, considering we were building a very large proportion of the world's shipbuilding output, was the extent to which the shipbuilding industry was a reflection of the behaviour of the UK as a whole. The product operated in a completely international market. There were no subsidies, and no concealed benefits. As far back as 1938, Sir Amos Ayre was talking about Japan as a long-term threat. I believe if you look at the history of Japan over the last 150 years, you have a country that set out on the military road to some form of world, or certainly, Pacific domination, to put it on a par with the white European nations. When it lost the war, it decided to pursue the same objectives through the economic route. They have achieved this with remarkable success. If people compare shipbuilding in the UK unfavourably with Japan, I would tell them not to forget television sets, radios, motor bikes and cars. It is the UK not being competitive, not simply the shipbuilding industry.

Dealing specifically with BSRA, I give credit there to the shipbuilding industry, because I think BSRA was a major success story. It was the first such research association formed by any shipbuilding nation. Britain was first for once [*Britain had also been the first country to truly industrialise during the Industrial Revolution, and had been first in devising a reliable steam engine through the work of James Watt on condensers, and for pioneering work on television, radio, cycles, motor bikes, radar, and jet engines*]. Its success was such that nearly every other European nation started setting up Chinese copies of BSRA. Throughout its whole existence, BSRA had great support from the industry in bad times and in good. I think the shipbuilders had a very enlightened outlook on matters such as international co-operation. They were willing that we should take the lead in setting up a scene for European co-operation.

The weakness in the BSRA support for industry was that for many years we made virtually no attempt at all to try and help firms with the application of our results on research. They believed that they competed with each other, and one of the things they would compete on was the application of BSRA results. In the same way they looked at production research, by saying that if we did production research then we would go round and see how various firms would do things. You would decide who was best, and then tell the others. "As my firm was the best," and everyone in the individual firms believed that – "we had everything to lose and nothing to gain." It was not until the industry accepted the fact that the competition was international, and some of those attitudes broke down that BSRA was able to be more positive in getting in at the development end, and to assist in the application of results.

It took the industry a long time to learn the lessons of the US Emergency Shipbuilding Programme. There, they would teach a woman to do a particular piece of welding in six weeks, and she then did that for thirty or fifty ships, and could not have taken you to the ship and showed you where her welding work was. We were still taking five years to train welders by way of apprenticeship.

One of the industry's strengths was the support that it got from BSRA, and the latter's tremendous prestige. International owners were very influenced by the fact that British shipbuilding had this body. In many ways I would say that they thought more highly of BSRA than perhaps the British shipbuilding industry did. A strength, which at times could prove a weakness, was the willingness to co-operate on a national basis. The weaknesses, I suppose, are that in the 1950s and 1960s the firms in the industry tended to be dominated by the single boss system. The big man, who may have been a family man, or someone who had broken through from the ranks, very much ran the thing himself. But do not forget the great days under that system. The industry was, however, becoming that much more sophisticated, and needed a change to the established system. That happened very gradually, with great difficulty and quite a bit of reluctance in the case of the industry.

The Patton Report [*Productivity and Research in Shipbuilding, 1962*] was done by the industry with a selection of some of the brighter young shipbuilders, Derek Kimber, Ross Belch etc. The new BSRA Head of Production, Peter Hodgkinson, who we had just appointed, was a member of the team. The report was largely about productivity, and I think the firms listened to them, and many were in a position to implement the ideas. It was a practical, solid working report. It also defined, in a sense, the first BSRA production research programme. Other reports criticism about the lack of investment is absolutely correct. The criticism about obsolete technology is only partially true; and it would be more correct to say that the UK was not in the forefront of the introduction of new technology. There was also no UK policy to deal with Japan Ltd. If a builder there had an idea for some new development that he wanted to see put into his ship, in Japan, the same conglomerate owned the shipping line, the shipyard, engine works, steel works, banks, it owned the lot. It could say "we will do all of this" and it could meet and justify the cost of the experiment. The Japanese government would actively encourage this, since they thought in terms of helping their shipbuilding industry. In Britain, the shipbuilder was expected to foot the bill for the lot. There was no co-operation from owners.

Prior to nationalisation, people were saying that the UK ought to go for the type of sophisticated ships with a high added value, and leave it to Japan to build giant tankers. During a debate, which Graham Day chaired, I challenged some of those present by asking how they thought that we were able to compete with the more sophisticated vessels, when we could not compete with the simplest. We had the capacity to design these ships, but why try and do it cheaper when we could not do a simple thing like a tanker cheaper.

You can look back and ask yourself where it all went wrong? I always look to the end of the war. We were heroes. We had stood [*mostly*] alone. Churchill had told us that this was our finest hour. My generation had seen their fathers decimated in the Great War and had been brought up to believe that that had been the war to end all wars. There was a certain sort of cynicism abroad. The Socialists were saying that they were going to build a land fit for heroes, and that they were going to look after people from the cradle to the grave. I think we began to breed people with an outlook that said the world owes us a living, and that we were entitled to the best. The State would provide, whereas, the Germans, the Japanese and the Italians had to rebuild their shattered economies; largely by their own efforts, but with Marshall Aid [*the European Recovery Programme, and in the case of Japan, through huge American investment to promote Japan as a bulwark against communism*]. You sometimes hear criticism of Marshall Aid to the Germans, but that begs the question that, as I understand it, we enjoyed far more Marshall Aid than the Germans did! However, we expended vast chunks of it in establishing a Welfare State. In a way that is probably why Churchill and the Conservative Party was rejected, because he would have said that we had won the war but that we still had to win the peace. It was not over yet, we had to tighten our belt. Indeed, it is only during the Margaret Thatcher years that that sort of attitude and outlook has been shattered to a considerable extent, but it is a pity that it has been replaced by "pull up the ladder Jack." In general, we took a wrong road and paid for it.

Returning to BSRA, we were extremely well supported by the government. We got regular grant aid [*around a third of what the shipbuilding industry was expected to provide*]. The only problem was that every so many years they changed the rules as to how you got it, and what you did to comply with it; and it got a little tiresome, but I was of the philosophy that if they made the rules then I would work to them. I think that what we wanted, and what was needed was a shipbuilding policy. Does the nation want a major shipbuilding industry?

Nationalisation was entirely a political decision. The tragedy of it is that it was an alternative to actually having a policy that might enable shipbuilding to survive, whether it was a subsidy to building, or a subsidy to owning, or better tax concessions or whatever. All these were things that could have led to gaining orders. Nationalisation was not going to do any of these things. However, it certainly provided a reason for pumping in public funds as subsidy, because, "it is now our industry," whereas you could not possibly do it when it was private, so the argument goes. The only time I felt that BSRA was left out of the industry, rather than being a part of it, was during the debates over nationalisation. We learned to stay neutral in the best interests of BSRA, and could not join in the industry's defence against nationalisation, or in defence of it. Including BSRA, in any event, would have raised the issue of hybridity, so we could not be nationalised.

I think that if one looks at the whole history of nationalised industries, one has to say that we are yet to find the form of it that works. Possibly because it removes a sense of responsibility from individuals it will never work; motivation is of overriding importance. Going through my days in the industry since 1935, I would not change anything if I had my life all over again. I suppose that one has to say that as far as I had a share in this, it is a share in failure. But, I believe that it is a failure, which is not attributable to the shipbuilding industry itself, but it is a failure nationally.

Interview No. 35: G.R. Snaith, Naval Architect, Head of Marine Systems Division, British Ship Research Association, Head of Headquarters Research Group, British Shipbuilders Plc

I graduated in naval architecture in 1952, and then worked in various design consultancies until 1964 when I joined BSRA with the brief to build up and lead a team designing standard ships for the then private shipbuilding industry, by voluntary co-operation among the companies. In 1969 I started a new division within BSRA called the Marine Systems Division. On nationalisation I was invited by the Chairman, Admiral Griffin to join British Shipbuilders to be responsible for the Headquarters Research Group. By October 1981, I became Production and Systems Advisor to the small ships sector within British Shipbuilders.

My first exercise on joining BSRA was on designing standard bulk carriers; the project was steered by a committee of fourteen managing directors of British shipyards chaired by Jim Lenaghan of Fairfield. We were beginning to feel the impact of Japanese competition. There was a widespread belief in the committee that the Japanese were cheating by the under-provision of materials in their ships. I already had experience of Japanese bulk carrier design and construction, and was able to produce evidence that on the contrary they were putting more in. They were rationalising the variety of scantlings on steel work and thus making it easier. When I started the Marine Systems Division at BSRA the concept involved marine engineering and marine transport systems. The idea was to take a systems approach to things. There was a terrible confusion in people's minds which took years to eradicate that they could not see the difference between engineering systems and systems engineering. In a nutshell, what I was trying to do was to apply systems engineering approaches to engineering systems. That palindromic statement is the nub of the misunderstandings which took place.

We did undertake certain major exercises including one into advanced automation systems. A lot of that technology has not yet been implemented, especially in this country. It was based on the emerging electronic technology of the day. We also looked at advanced interactive computer aided design methods, focussing on ship structural design. The important thing was that we were too far ahead of the ability of the computing industry to support it. I went to British Shipbuilders, but after a year I stopped the project because the infrastructure technology from the computer industry was not there. I was responsible for the Research Group, not research and development. In my experience over many years, the term research and development is glibly tossed around without being thoroughly understood. I see the output of research as being the structured knowledge base, which is then used in the context of either product design when you are using that knowledge to eliminate the technological uncertainties of the product, or to use the process design.

The British shipbuilding industry was dominated by a craft approach to construction. There is one particular feature of this, which does not deprecate the men themselves whatsoever. As soon as you are committed to a parts and assembly production system that involves rework, you are already committed to a vast number of extra man hours. This was part and parcel of the way of doing things. Accumulated geometric errors were made in the stages of the steel assembly work, so you got to a stage where assembly on the berth was a monumental exercise in pushing and squeezing huge chunks of steel to get them to mesh and lock together. I can show you evidence from small to large yards, which showed the clamping arrangements, the bridging pieces that would be temporarily welded on, and so on. These are fundamental weaknesses in the method of carrying out the work. Associated with this was the apparent inability to design out needless work, and a lot of the latter was actually designed into the job. There were also major problems with materials control.

Poor management was really concerned with the organisation of work. Also what constitutes control, and what is the role of management? You find numerous examples which amount to no more than a description of fire fighting, which is a sort of macho approach to management where people who are undoubtedly skilled like to have situations which require the input of their John Wayne type of authority. It is bombastic, directive, dogmatic and pontifical, but it has low learning. It does not lead to the avoidance of problems. Shipyards are intrinsically complex. Their products are complex. Therefore, there was a widespread belief that they could be understood only broadly. In modern parlance there was no structured set of models of how the thing was perceived. If you talk about models to the Derek Kimber's of this world, they think that it is nonsense. They have not realised that everyone is a modeller. There are only two types. There are those who know what they are modelling, that is, creating simplified representations for purposes of description and prediction, and there are those who do not. Those who do not know that they are modelling are dangerous.

Capital investment may be necessary, but of itself, it is not sufficient, you must have the avoidance of the waste of time. This does not mean as it has been interpreted in some instances that you require macho management cracking whips and making people work harder. The trick is to work smarter. I recall going with a party of British shipbuilders to visit the Mitsui shipyard in Japan. When we were walking around the yard in the morning, we went into one of the assembly shops, and it seemed as if it was lunchtime. If you looked closely you could see the workers, but there was no feverish activity. The important thing was to work consistently to high standards of performance, and get your output right first time. That requires the whole thing to be structured and organised by people beyond the craftsmen themselves.

The ships that were simple, both geometrically and technologically were bulk carriers and two deck freighters, SD14 type vessels. If you looked at ships which were geometrically complicated but technologically not sophisticated you would get cross channel ferries. If you looked at both geometric complexity and technological sophistication then you had gas carriers and chemical carriers. The whole thrust of British Shipbuilders marketing efforts aided, led and abetted by T. John Parker, was very heavily into the merchant side of simple geometric and simple technological vessels. This was low added value stuff, and it did not require a great deal of technological sophistication.

What is required to deal with the more complex vessels? Geometric complexity requires you to have good organisational management, and good organisation of assembly. If it is technologically sophisticated it requires advanced technological knowledge in order to design the onboard engineering systems. You need also the skills of assembly of complex vessels. As I perceived it, the industry was walking away from that complexity. There was a pathological wish to avoid difficulties by avoiding complex vessels. It is my view that management on so-called secondment from certain shipyards to British Shipbuilders Headquarters were protecting the continued independence of their individual companies, and were therefore against central product development. They argued, quite foolishly in my view, that owners wanted vessels to be built in particular yards. There was never any real willingness to pool resources.

One thing that did not change with nationalisation was precisely the weakness of management. They were the same people at all levels. There was a continuity of incompetence. There was no injection of know-how. I went to Hall Russell to give a seminar on Japanese methods, and the Managing Director said that it did not apply to them. That was a ruinous attitude, and was totally and utterly wrong. Basically, management was largely under-educated, and certainly under-trained, and put into management roles without any instruction. Their method of working was idiosyncratic to them as individuals. It was probably modelled on who their previous boss had been. They were enormously confused between authority and their competence. There are two facets to this. In one sense they pontificated authoritatively. They wielded authority. The other side of that was the sort of authority exercised by Ross Belch [*Scott Lithgow*] where nobody was right except for him on everything. Of course, he was generally wrong.

They did not adequately understand how to define people's roles, what they should be doing. It required an overall vision of how the shipyard should operate at every facet, whether that was executing design, or whether it was in the way that work was organised and executed.

You will not find in the Geddes Report, for instance, anything which shows people how to be better in terms of ship construction. When in Geddes they talk about products and markets, there was an ironic thing taking place while Geddes was pontificating to British shipbuilding about products, his own company, Dunlop, was almost destroyed by being overtaken by steel reinforced tyres. All the reports into British shipbuilding failed because they did not give the answer to the question of how to do it. Simply to amalgamate people with the same low level of competence did not achieve anything.

The industry could not have survived without nationalisation. Prior to it, Admiral Griffin and Graham Day went round the companies [*before Day resigned in December 1976*] and the key message that was reported to me was, "do not worry, you will not hear much from us, just get on with things." That attitude, which the companies of course welcomed, was carried over into nationalisation, so that there was on the ground resistance from every company. The attitude was that anything for HQ was pernicious and should therefore be resisted. Because of accountability to the Government there had to be a centralised financing and corporate planning structure set up.

The major advantages tended to be lost opportunities. The advantages were centralised marketing, centralised and co-ordinated design, and treating the individual yards as cost centres. These opportunities were lost. With the warship yards the situation was somewhat different. They were most exclusively dependant on the Ministry of Defence for work which was certainly not on a loss making basis. Companies could not lose on a project. This led to an air of what I regarded as needless arrogance on the part of the warship firms. They had the idea that they were right and the merchant builders were wrong. In fact, they were all incompetent.

Overall, there was a basic blindness to understanding that there was a fundamentally different way of assembling ships that was inherently cheaper. It would have made the ships competitive. Allied to that was a structured approach to marketing and the design of technologically smaller ships. There was no vision, because people were confused by the apparent complexity. They were confused by the kaleidoscope of geometry, of product geometry, of organisation of facilities. They did not have an adequate set of models to explain that complexity.

Interviews
The Trade Unions

By the time of the 1965-1966 Shipbuilding Inquiry Report, the vast majority of manual employees in the British shipbuilding industry were members of thirteen trade unions, eleven of which were craft-based. In the larger shipbuilding firms, union membership was 100 percent owing to the "closed shop" system where union membership was compulsory. The majority of workers engaged in hull construction belonged to the Amalgamated Society of Boilermakers, Blacksmiths and Hull Constructors. This union arose through a series of amalgamations in order to minimize demarcation disputes among different unions, especially between shipwrights and platers, and to present a more united front to employers. The other unions were mostly engaged in the fitting-out trades. All shipyard trade unions were affiliated to the Confederation of Shipbuilding and Engineering Unions (Confed) which acted on behalf of its constituent unions in national negotiations on pay and conditions common to the industry, but crucially it had no powers to negotiate on matters specific to a particular craft, including demarcation and wage differentials. At the yard level, each trade had its individual shop stewards, who in turn belonged to a Shop Stewards Committee under a Yard Convenor. Apprenticeships were uniform in the skilled trades and usually lasted five years. Union membership was less rigorous in the more casualized ship repair sector, but it was general in marine engineering where the majority of workers belonged to the Amalgamated Engineering Union, which it turn was affiliated to the Confed.

Interview No. 36: James Airlie, Convenor of Shop Stewards, UCS, National Official Amalgamated Engineering Union

I served my apprenticeship as a fitter with Lobnitz and Co., at Renfrew who built dredgers. After my National Service, I did not stay long after returning, and went to work in other industries, before returning to shipbuilding to work at Fairfield in 1965, just before the firm collapsed. I was an active shop steward throughout UCS, Convenor, full-time official, and I am now a national official on the executive of the AEU.

Belatedly, Fairfield had spent a lot of money trying to modernise, although subsequent events were to show that the money had not been well spent. The old management were not up to managing a modern competitive company. When the industry began to face the growing competition from the Japanese, the Scandinavians, and the Germans, they were ill equipped to face it. The more intelligent elements within the shipbuilding industry, and I use the term "intelligent," advisedly, moved the money into more profitable industries.

In many ways the decline of the British shipbuilding industry was a microcosm of the decline of Britain as a major manufacturing country. You could see that in a number of ways. It was badly run, and was badly managed, and the employees were badly motivated. In other words, it was a catalogue of inefficiency and stupidity. My view is that it was basically down to the employers. That is not to say that there were not working practices and demarcation disputes, and tribalism from the side of the unions. But, basically the motivator for any constructive change in any industry, in my opinion, must be the management. Trade unions are not equipped to fulfil that role, though they can help the management in such a change.

They tell the story that when Fairfield collapsed, the Government brought this new consultant in to advise how they could rescue the yard prior to being taken over by Iain Stewart. He asked one of the managers if they operated critical path analyses, and he though he was talking about a Greek super tanker! So they did not have any basic management skills, or understanding. It was a backward industry, tribal, small-scale in many places. There were no strategic objectives for the industry. Britain did not see the need for a strategic overview of the economy, particularly manufacturing. The basis of the criticism by Geddes of the small-scale of the industry was spot on.

It is true to say that the move to grouping was not carried out properly. The restructuring was under-funded. It had to be underwritten, in my opinion, with public money. Secondly, a clear view of the market needs and trends was never accurately analysed. Thirdly, the rationalisation which should have taken place on the back of restructuring never took place. In place were the overall managers within the yards, so there was internecine warfare amongst the management. There were petty jealousies, and lack of control and cohesion. It took a number of years to begin to tackle these things. Losses mounted and productivity did not increase to the level that was required to make the industry competitive. Subsequent events have made me think that we might have been too optimistic to think that a developed nation could have become competitive in shipbuilding. One of the tragedies of shipbuilding was that it was a very important industry in employment terms. It was primarily located in areas where there was higher than average unemployment, in many ways, therefore, the subsequent collapse of the industry was a disaster.

UCS was not sufficiently funded. There were a number of promises given in the period before the election of the Conservative government in 1970. It was clear that there were major problems with UCS, and that there needed to be a further reconstruction. A tri-partite agreement was arrived at between the government, management and the unions. The management was committed to do certain specific things. The unions would accept a degree of rationalisation, efficiency and relaxation of demarcation. The government, for its part, was to provide more funding. That was the agreement, but the Conservative Party won the election in 1970. The Tories [*Conservative Party*] had met at the famous Selsdon Park Hotel Conference when the concept of "Selsdon Man" was developed. One of the architects of that was Nicholas Ridley, a right-wing ideologue. They developed the Lame Ducks philosophy that the State would no longer bail out failing industries. UCS needed further money to meet its short-term needs, but in the long-term it needed more and more money. The Government kept UCS dangling because they had faced the collapse of Rolls Royce just beforehand. Once they resolved that, they reneged on UCS and called in the liquidators.

This was all premeditated. We obtained a copy of an internal memo made by Ridley, where he spelt out the strategy. Yarrow was to be extracted from UCS, and the whole of the rest was to be let go. Anything that could be rescued commercially should be tagged on to the lower reaches [*of the Clyde at Scott Lithgow*]. That was the Ridley Plan. It was one of our most important tactical and political coups to get a copy of that memo, which we then leaked to the media. The Tories were rocked right back on their heels.

The unions took a clear decision that we were going to fight this for a number of reasons. The demand for ships was booming at that period. Basically, the government were vulnerable, in our view to a well conducted campaign. Shipbuilding then played such an important part in the economy of the West of Scotland. Politically, most adults either worked in the yards or had a relation who did. No family was unaffected by shipbuilding. We decided that going on strike was of limited value, so we decided to confront them with a major tactical problem by taking over the yard. We rang rings round them politically for fifteen months. We won, basically, all the demands. Govan Shipbuilders was set up [*with yet another huge cash injection*] including Scotstoun and Linthouse originally. A buyer, Marathon, was found for Clydebank.

It was important, as far as I am concerned for getting the Labour Party to commit itself to bring the industry into public ownership. That was an important by-product. I think we not only influenced the Labour Party, but we were instrumental through our campaign to adopt nationalisation as a policy. Subsequently, events did not work out as planned. In the first place the crisis in shipbuilding was deeper, and confined not just to Britain. Nationalisation could not have come at a worse time when the demand for ships was seriously declining.

I recall when UCS collapsed there were a number of pundits arguing that the strategy should not be to save UCS, but to build new yards on Greenfield sites on deep water, thus firms would not be limited on the size of vessels that could then be built. That was the accepted view of some of the economic analysts. We argued that it was nonsense, for two reasons. The trend of the world trade was between the developed world and the developing world and the underdeveloped world. Roughly eighty percent of the harbours of the world were incapable of handling vessels of over 40,000 tons deadweight. The second argument was that very large crude carriers were fine if they were economic, providing there was a full cargo. If there was a major recession then there would be no demand to fill a large crude carrier, and this type of ship would therefore become uneconomic to build or operate. We have been proved right in both cases. The demand is now for smaller vessels, but more flexible and technologically adaptable. The point I am making is that the experts are not so expert.

Most of the managers in the industry became so for two different reasons. One was that they had knowledge of shipbuilding and were technically competent. The other ones were perhaps offspring of the old owners and were not very clever. I will deal with the ones that did know the industry. I would not dispute that they were very knowledgeable people. It does not mean that you necessarily know how to manage, how to develop a strategic overview. That is what the industry lacked. It would be a mistake to say that they were not good people. By and large, shipbuilders were a macho type of people who had the attitude that no one could tell them how to build ships, because they had worked there all their lives. That was the accepted view. The old type believed in the divine right of king's. That is not to say that the unions did not bear their responsibility. There was tribalism, and sectionalism; working practices were guarded zealously. The Boilermakers were particularly involved. I have no doubt that it did contribute to the decline of the industry. Unions were accused of being too powerful, but in opinion we were not powerful enough. A union, by itself, can not bring about fundamental change that is not the basis of its work. There has to be management that poses the challenge to you, and gives you an explanation. People are not unintelligent. Management never operated like that. There was no way that fundamental change would occur unless the management changed. It is true that certain changes did take place, and it is true that a new view developed. My overall view is that if an industry goes into decline, then the responsibility lies with the people who own it, manage it and run it.

I think we can sometimes be too critical of government, but there was no strategic view of where we were going as a nation. Do we want a shipbuilding industry and are we prepared to meet the cost. The delay in the run up to nationalisation was extremely harmful. On nationalisation, I think we needed a tougher regime. That did not happen. Who knows? Perhaps it was the case that we had to take people along gradually. There has to be a learning curve until the lie of the land is known. The problem was that there was not enough time. The Tories came in and decided to get shot of the industry, to let it wither on the vine, with the exception of the warship yards.

I do not think that the Tory policy to let the industry wither was their philosophy, but they did decide not to pour any more public money into it. If it survived, then fair enough. They had a non intervention policy. I recall Teddy Taylor [*Conservative MP for Cathcart in Glasgow, and later for Southend in Essex*] saying that it would be cheaper dropping £10 pound notes onto the workforce. It is a naïve, stupid concept. You should be talking about meaningful employment in an industry which has a future. And the industry has a future because there is no better way of transporting in bulk over distance.

Under British Shipbuilders industrial relations improved as there was close consultation at the national level. There was a recognition that you had to take the workforce with you if there were to be fundamental changes. By this stage the unions recognised that the market place was tough. There was also a redundancy package which made it more palatable to a number of people. That is not to say there was not concern, but there was recognition that restructuring had to take place. What was completely criminal was the way the Tories washed their hands of the industry. In the future, when the history of the 1980s is written, the Thatcher Government will be seen as the most disastrous government this century. They have done great damage to British industry. Shipbuilding was symptomatic. They introduced privatisation to get rid of the industry. That was not a dogma. They just gave it away with the exception of the naval yards, and for political and strategic reasons Harland and Wolff in Northern Ireland.

Looking back, I think the 1960s was the key decade. If the Labour Government had had the foresight then, they could have used Geddes as a base to nationalise the industry and proceeded to restructure it by making the finances available. There should have been more decisive action then.

Interview No. 37: Jim McFall, General Secretary of the Boilermakers Society

I started work in Harland and Wolff in the late 1940s as an apprentice plater. I then became involved on the trade union side and was then elected to the Executive of the Amalgamated Society of Boilermakers. Eight years later I was elected General Secretary, a position which I hold to the present day.

In the old Boilermakers situation there were seven trades. Each one bordered on the principle of not any one of the other six could partake in the trade of another. There had been no investment in the industry for many, many years. I think the lack of investment was a greater difficulty than the lack of change. The one thing that an increase in investment does, bringing in an increase in the use of high technology, is to in itself stimulate change. The new equipment demands new ways of working. The standards are lifted dramatically. It helps bring about change in the trade union movement. The shipbuilding industry was one of the industries that never had the investment even in the good times. I worked in Harland and Wolff during the good times when they were building all sorts of ships from aircraft carriers to bulk carriers to passenger liners. They were making capital and were not putting anything in. You would work with machinery that had been there for fifty years.

You must create the difference between the imagination of a group of workers, and the imagination of a group of employers. As competition was becoming more severe, because the style of ships were changing, and the market place was getting wider we had a multiplicity of shipbuilders. At no time did I see those shipbuilders say, "let us come together as a consortium." The old shipbuilding people were a breed of people to themselves. They were families. The Rebbeck's and the like sincerely believed that they were the saviours of society and did not need any assistance. Maybe the Government could help by giving them a knighthood, but they did not need anything else. That was their view of life. Unfortunately, the people who were emerging in the developing countries looked at industry very differently. They examined it from the starting point of what the industry required. By contrast, we struggled on with no commercial instinct whatsoever. All the things they blamed the trade unions for could be turned round on them. They would ask, "why should there be a multiplicity of trade unions?" You can equally ask why should there be a multiplicity of employers.

Before nationalisation the weakness was that the industry was totally disintegrated. There were too few ship orders in British yards with the advancement, barring Japan, of other countries which had not built ships before. Korea is the best example, but there was also Brazil. Closer to home was the competition of Germany, Spain and Portugal. I felt that the co-ordination of the UK industry was poor in itself. West Germany was much better organised as an industry about to go through some traumatic experiences. We were aware of the German unity between the employers, the trade unions, and the banking and commercial system.

I felt that nationalisation could have been better managed. I think that genuine mistakes were made. There was huge opposition to nationalisation from within the industry that remained from vesting day in July 1977, to privatisation from 1984 onwards, and the opposition was particularly strong from the warship firms. They felt that they were the most important people in the industry and that they would deal with it in whatever manner they liked. That was the feature that always came forward from Vickers and to a lesser extent from Yarrow. They never saw themselves as part of a nationalised corporation, but saw themselves as part of the instrument, shall we say, of the defence of the realm. Some of the merchant builders did not accept nationalisation in principle as well. There were few people in charge of British Shipbuilders who followed the pursuance of nationalisation.

The advantages for the nationalised industry were brought about by the trade union movement. We had a difficult task. We were bringing together twenty or more yards with the complexity of the associated demarcation structure. We then developed one negotiating posture, which is easy to say, but very difficult to manage. We brought about in 1977 the most aggressive form of national negotiations ever seen in this country, where we wielded the power and influence in order to try and protect the industry, and to overcome the divisive nature that we inherited. We did that under out past general secretary, John Chalmers. He felt that if a corporation was nationalised by a Labour Government, then the people within the Labour movement should do their best to bring about prosperity in that industry. In my opinion, he brought about a Shipbuilding National Agreement that was *par excellence*.

We tried to separate the two entities in shipbuilding; the steelwork and the outfitting. This was never terminology prior to nationalisation. We fought to establish the principle that within steel working as a whole anyone could do anything within the seven trades. There was a bitter battle within our organisation. It was the hardest battle because our members perceived that they had a lot to lose and little to gain. The agreement was established in 1978, although we progressively moved into that sphere.

The National Agreement was very difficult to implement. One of the most memorable periods was the Polish order when we got, if my memory serves me well, about thirty-seven ships. They were brought in to an industry starved of orders. There were no orders in Swan Hunter, Govan, or in NESL in Sunderland. So there was a re-distribution process brought about. There was an upsurge of hostility in Swan Hunter over a flexibility agreement, between the then boilermakers and the outfit trades. It could have not happened at a worse time because British Shipbuilders was struggling for survival. The union, nationally, went into Swan Hunter and expressed the point of view that these things of the past should be put behind us. We should move on to create the conditions for the survival of the industry. The Swan Hunter shop stewards totally disagreed. Not one was in support of anything other than the demarcation issue of the day.

As a result we then had the Govan Shipbuilders situation. It was paramount to them that they had fought during the time of UCS for the retention of shipbuilding on the Clyde. Here they could see that another area other than the Clyde could eventually bring about the destruction of the whole industry. The Govan shop stewards said that they would build any ships that would not be built on the Tyne. Jimmy Airlie was the Convenor then. That then brought Swan Hunter to the point where they had to take a decision. The Polish order was a breathing space to ensure a recoverable position for the industry, safeguarding it until British Shipbuilders could get their marketing team together.

The Labour Government did not negotiate the Polish order but announced it in advance. British Shipbuilders built these ships at great expense to themselves that the asset was the maintenance of the merchant building sector.

When we negotiated with the government over the Polish order, we came to the understanding that there were too many people in the industry. We then argued the principle that these people had served the industry for their lifetimes, and therefore there should be some severance payment for these people who were leaving the industry through no fault of their own. The term, redundancy, was not used. That resulted in immediate negotiations under Gerald Kaufman, a junior minister. This led to the inception of the Shipbuilding Redundancy Payment Scheme. It was reasonable, if not as generous as we would have liked. It was a good opportunity to look at the age profile, and we insisted that those people who were to be given the opportunity were to be between the ages of sixty and sixty-five, if they wanted to go. Here was something that shipbuilding workers never had in their lives. They had always suffered the pains of unemployment, the product of a cyclical industry. Here was an opportunity for people to get £3000 to £4000 as payment for service. It was an incentive for people in that age range to retire. On top of that the Boilermakers had already begun to negotiate payments for industrial deafness. So a man might get another £3000 to £4000 on top of his redundancy payment. These were the compensatory factors.

After 1979 the situation changed, and the Tory Government imposed a limit to the involvement of the trade unions. Secondly, the preparations for privatisation were put in hand. It was never openly said. However, the naval side were happy to return to their old domain.

Interview No. 38: Alex Ferry, District Secretary, Divisional Organiser, Amalgamated Engineering Union, Glasgow, General Secretary of the Confederation of Shipbuilding and Engineering Unions

I became steadily involved with shipbuilding at the time of the UCS Work-in, and from that time onwards remained heavily involved; and in 1978, I became General Secretary of the Confederation [*known as the Confed*]. I was right in at the beginning of all the work which started with the nationalisation of the industry.

The weakness of the industry was complacency on everyone's part. The managements failed to invest, and there was a lack of forward thinking on their part. Frankly, the trade unions lived a misconception, because we told ourselves that we were part of an efficient, highly productive workforce, when in essence our productivity was rock bottom. Our working practices were outdated, for understandable reasons, maybe, but they were still outdated. Our work rate was absolutely chaotic. The trade union structure in the industry gave rise to all sorts of demarcation problems and industrial disputes. We were not very clever. We were complacent all round. It was the structure, history, and traditions. If you went into ship repair, it was ten times worse. Ship repair was a casualised industry for a long time. It was not until about 1969, that we established a payment for the workforce when there was no work available. That was established at the Elderslie Dry Dock. In ship repair, the ships were queuing up to come in, the management were doing deals with the captains, and vice versa, and our members knew what was happening and put the screw on the management. You could get treble time in wages for doing certain jobs. I have to say that ship repair was almost corrupt in every sense.

There was never a policy for shipbuilding. If you take nationalisation in 1977 then you can say there was a policy, a direction. There was a hope. There never has been under any government an integrated maritime policy. That is why we have seen our shipbuilding and shipping interest decline. We have had no maritime strategy at all. If a country has a government who projects a philosophy and a policy of partnership such as in Japan, where the government, industry and the banks collaborate in order to create industries, and build a very strong domestic market in order to leapfrog into the world market with a competitive edge, that is not unfair. That is a thought out policy, which, up until now has paid dividends. We have never had that. Even with nationalisation some of the owners and senior management of the yards that were nationalised, sabotaged it. We never created the spirit of trust.

Conservative governments have always had a desire to keep themselves at very long arm's length from industry. The lame duck philosophy was a typical example of that; if you can not survive on your own then do not look to us for help. That did a tremendous amount of damage, not only to the shipbuilding industry but to other sectors as well. Quite frankly, the Thatcher governments pursued the lame ducks philosophy under a different guise and went much further. They pursued the policy of putting the public sector into private ownership. The fact that they cost the taxpayer millions in the process did not matter; they were getting rid of it that is what they think.

Labour governments have acted differently, but I have to say that they never really had an economic plan for the industry which proved to be workable. The will has always been there. I would have to question whether the billions of pounds poured into the industry actually gave value for money. The industries that the money was being poured into were still under the same management and ownership with a few people put in here and there to look after the government's interest. The managers, and I am talking about the top people, were able to take advantage of the situation. They did all right for themselves. They were taking all the money that the Labour Government could give them, but at the same time they were doing all they could to undermine the government and its policies. They were living for the day when they would get the yards back into their ownership.

On nationalisation there was a tremendous enthusiasm. The hours we put in were unbelievable. The Polish deal was about buying work, without it the industry would have been dead there and then. There was commitment from government. There was the response from the trade unions. British Shipbuilders were, and still are, the only national company to have ever succeeded in negotiating a single agreement to cover every employee. That was a remarkable achievement. The barriers on flexibility, demarcation and interchangeability were broken down. A practical plan was laid down where we accepted as trade unions a very substantial reduction in capacity. British Shipbuilders management were saying; "below this capacity we can not fall or we do not have a shipbuilding industry."

The disadvantages were that most of the managements were not injecting the same degree of enthusiasm. Our members, while they were responding to change, also had the feeling that they were now in a secure position. The one thing we never conquered at British Shipbuilders, despite all the efforts, was the productivity problem. We had internal sabotage. We had a misunderstanding on the part of our members in relation to the job security element. Also, although we had commitment at Board level, it is doubtful whether we had competence. There were strengths and there were weaknesses, although it has to be said that there were saboteurs on the Board. There were some who had absolutely no commitment either to the industry, or to the principle of nationalisation. There did not appear to be any control on expenditure. We did pull it back considerably, but there were disagreements between Board members and the yards. The basis and the idea were sound, but we never had long enough to make it work.

Interviews
The Civil Servants, Board of Trade: Shipbuilding Inquiry Committee, Shipbuilding Industry Board, Ministry of Technology, Department of Trade and Industry, Department of Industry

Government supervision of the industry after 1950 rested in a number of departments. Up to 1964, responsibility for those firms engaged in warship building for the Royal Navy rested in the Admiralty. Thereafter, on the abolition of the Admiralty, responsibility for this sector was transferred to the new Ministry of Defence, where it still remains. On the mercantile side of the industry, overall responsibility lay first with the Ministry of Transport (1939-1964), Board of Trade (to 1966), Ministry of Technology (to 1970), Department of Trade and Industry (to 1974), Department of Industry (to 1983), and Department of Trade and Industry (to 2007). Current responsibility for what is left of the mercantile sector now lies with the Department of Business, Innovation and Skills.

The Shipbuilding Inquiry Committee 1965-1966 was appointed by the President of the Board of Trade, Douglas Jay, and was chaired by Anthony Reay MacKay Geddes of the Dunlop Rubber Company. Its terms of reference were to establish the changes that were necessary in organization, methods of production and other factors affecting costs to make the shipbuilding industry competitive in world markets; to establish the changes in organization and methods of production that would reduce the costs of manufacture of large main engines of ships to the lowest level; and to recommend the actions that should be taken by employers, trade unions and government to bring about these changes.

The Shipbuilding Industry Board (SIB) was set up as a result of the Shipbuilding Inquiry Committee 1965-1966. The recommendations of the Committee included a programme of assistance for the industry through grants and loans from public funds for a limited period of five years, to be administered by a small independent board. On 9 August 1966, the government announced its decision to implement the report and to introduce the necessary legislation. Initially, the SIB was set up on a non-statutory basis with three members; in accordance with government wishes, no one with previous experience in shipbuilding was appointed. The Chairman, Sir William Swallow, came from Vauxhall Motors, Anthony Hepper from the Thomas Tilling Group of companies and Joe Gormley from the National Union of Mineworkers. In November 1966, the Ministry of Technology assumed responsibility for the ship-

building industry, and in 1967 the Shipbuilding Industry Act was passed, establishing the SIB as an independent statutory body but subject to the general direction of the Minister of Technology and answerable through him to Parliament. The Act established the SIB for a limited period, initially up to the end of 1970, with provision for an extension of up to one year. The SIB was eventually dissolved on 31 December 1971.

Interview No. 39: Reay Geddes, Chairman, Shipbuilding Inquiry Committee, 1965-1966

I was chosen to head the Committee because I had no prejudices about the industry at all. They wanted a general purpose industrialist. The nature of my involvement was simply that report. There was nothing either before or after. The government had decided to have a Shipbuilding Industry Board, and it seemed to me that I had no role anymore. I declined to answer anybody's questions after that because I did not want to cause any confusion. Journalists love to find differences of nuance. I stepped right out of the industry.

I was not particularly keen to do it as I had no special knowledge, but I was willing if Anthony Burney could be the accountant in the team. I knew him well as a friend, and I also knew the quality of his work. The Board of Trade, the sponsoring department, suggested the other members of the Committee for obvious reasons. Fraser was suggested as an engineer; Professor Lighthill as a senior academic research scientist, Scott as a management consultant, Charles Smith as a trade unionist [*General Secretary of the Post Office Engineering Union*]. Professor Thomas Wilson came from Glasgow University as an economist. Then we were given Victor Chapman as Secretary, who was a very able civil servant. He had been Private Secretary to Harold Wilson, so he was a high flier intellectually. We had one or two meetings where we decided how to go about it, and discussed what was expected by Douglas Jay [*President of the Board of Trade*] and other ministers. We then worked out the sequence in which we would do things, both in terms of visits, and in terms of hearing chosen witnesses [*written evidence was publicly asked for*]. We then started the detective work. We included big customers as well as suppliers like the steel industry as important people to question. We cast our net fairly wide.

We had some difficulty in meeting the trade unions. There was a great personality in the industry called Danny McGarvey [*a Scot*] of the Boilermakers. He would not meet us. We kept offering him dates in Newcastle and in Glasgow. Finally he agreed to come with some of his colleagues and have a meal with us in a pub in Newcastle one evening. We assembled in good time before dinner and chatted. It did not seem to be a meeting of minds at all until he said, "come on lads." So they all got up and walked out. I said to the others, "sit tight, I do not know what this means." We sat there and he came back in. He had changed totally. He said that they had never had a meal with boss types like us before, and they expected that we would put all the blame for the industry's ills on them, but we did not seem to be prejudiced and appeared to want the industry to do better, so they agreed to talk to us.

In the end, the government adopted the Report, which meant adopting the SIB, and the promise of help conditional upon rationalisation and modernisation plans worked out to the satisfaction of the Board. On grouping, if you took a company like Shell, which was a successful company, it was not afraid of having a staff of high quality people. That went for ICI [*Imperial Chemical Industries*], or any of the other big successful companies. The same went for the Japanese or of any of the big yards on the continent. They would have an adequate planning staff. It would not just be done by the owner and his apprentice son. We thought that the chance of getting a planned load of work, which would be properly conceived and marketed, would be greater with bigger entities than with the existing ones. If the group wanted to be both a warship builder and to specialise in bulk carriers for example, they could divide the work among different yards, but with the backing of the marketing and financial and buying strengths that came with a group. It simply was a question of the benefits of scale in a qualitative sense, as well as quantitative. We were in favour of greater added value specialists ships, because there was more skilled work that could go into them.

We were aware of a fair amount of enmity between yards in grouping. Yarrows were cross with us. They were quite sure that they were special, and they were. They were very good at what they were doing, but they were relying too much on naval orders.

I came from a process and mass production industry, which had continuous shift production. We also had a considerable engineering works in Coventry, part of which was of a jobbing character. It was never master of its own destiny. It is only when you have got strong enough research and marketing to be able to plan ahead in good times and in bad that you become master of your own destiny. In shipbuilding they would pride themselves that they could turn their hands to anything.

I have never felt that we drew the wrong conclusions from the information available to us. Everybody on the Inquiry worked very seriously. I failed to foresee the oil shocks of the 1970s, but I was not alone in that.

Interview No. 40: Victor Chapman, Head of Shipbuilding Branch, Board of Trade, Ministry of Technology, Department of Trade and Industry, Secretary to the Shipbuilding Inquiry Committee

I was having a sabbatical year in 1964, and towards the end I was called from it to take charge of the shipbuilding branch of what was then the Board of Trade. Almost immediately afterwards, the Shipbuilding Inquiry Committee was set up. I was Secretary of that with very full time involvement. That was the year of 1965, broadly speaking. Then, in 1966 I switched attention to my real job, as it were, as head of the shipbuilding branch. There was at this time the promotion of legislation and the setting up of the Shipbuilding Industry Board. I continued with that same job as it was moved about from the Board of Trade to the Ministry of Technology, and then back to the new Department of Trade and Industry. I was then posted to quite different duties in 1970.

One had to see the shipbuilding industry against the background of the general industrial psychology at the time. Looking back, we were stuck between two things. On the one hand, there was the late 1940s phrase, "the commanding heights of the economy," and in that context, shipbuilding came near to coal and steel. This was combined with the feeling that Britain was a seafaring nation that built ships to travel the world. On the other hand, one was already beginning to see that the shipbuilding industry might become, what is in the language of the 1980s, a "smokestack industry." We were between the two in the 1960s. Shipbuilding was seen to have its special problems.

When we looked at shipbuilding the first thing that hit us, and me particularly, was that it was an odd man out. In a sense it had no home market. One has to remember that a British ship owner could buy a ship from a British yard, and once it was fitted out and left the berth it would not necessarily come near this country. You were selling to an international industry, with one exception, and that was the peculiar exception of the Royal Navy.

As a Committee, we went round the yards, and we talked extensively to both sides of the industry. That is all set out in the Report. One learned a lot from that. There were two real gaps in the information. One was inherent, and is a concern of mine. It is an odd industry in the sense that it is really difficult to measure output in any simple way. Then there was the question of the market. There is a lot in the Geddes Report about the market. We got a good consultant in to advise us on the market. He came back to us after a few weeks and said that to do a proper econometric survey would take two years, and that we had allowed them just three months. To do the job properly it was necessary, he argued, to estimate the future of all the main loads all over the world. We decided we would have to make do with the methods of their mathematician. He was to make a pretty sophisticated extrapolation of trends. There were two major difficulties. Firstly, it was, with the exception of the navy, an international market. Secondly, it was a market that was peculiarly difficult to forecast. After all ships are very big capital investments and they suffer from the ups and downs of capital investments, perhaps more decisively than most things.

The preamble to the Report should never be overlooked. It was largely Geddes's own work, and he would ask, "can we put this sort of thing into a blue book?" Just as the origin of the Committee was of political concern, so I think that the problems were largely psychological. All the defined economic things mattered, but it was much more a question of psychology. All of what we now call the smokestack industries were lacking in confidence. I think shipbuilding was particularly lacking in confidence. They felt so squeezed between big suppliers and an international market. Both sides of the industry lacked confidence. It was this that forced them back to the attitudes with which they had grown up. The Report was really an exercise in confidence. We knew that an endless list of technical recommendations, all of which were necessary, by themselves were insufficient. It was Geddes particular, personal contribution to emphasise this element of confidence. You would never know this from his speeches, however.

On the Government side, the Report went through without any problems at all. There was fight in the Committee over financial terms. I, as Secretary, should have been neutral, but I was a protagonist. Tony Burney [*a highly respected financial accountant from the firm of Binder Hamlyn*] was keen to do a terrific thing on credit arrangements for home shipbuilding. My argument was very simple. I thought it would be fatal to the Report. Journalists would never put the word "credit" in the headline. The headline would have been so much money for the shipbuilding industry. That is why I eventually won Geddes over. We felt that we should recommend a relatively small amount of credit, plug in this two percent on home shipbuilding, and then give a certain sum of money for the recommended grouping of yards.

I actually wrote most of the Report, with three exceptions; the human relations piece I revised with a member of the Committee, Burney wrote some of the material on management consultant stuff, and Geddes wrote the preamble. It was very hectic. When completed, it went down beautifully. Sometime after there was a meeting of North East Labour MPs where one is quoted as saying that, "it won us the election" [*of 1966*] in that part of the world.

We sent out a popular version of the Report with the co-operation of the employers to every employee in the industry, which was the objective of the Government. The timetable was kept [*handily, just before a General Election*]. The Shipbuilding Industry Board [*SIB*] was set up according to plan. It really began to lose its way, I think. A second string running through this was that the Ministry of Defence (Navy) were involved. They wanted to maintain a big institution in Bath. They wanted competition and were terrified of falling into the hands of a single supplier. This knocked a lot of the stuffing out of the SIB, and was one of the reasons why the grouping of yards was not successful. This was one of the major problems, which perhaps do not come out because it was an internal government affair.

The international scene was very odd. I refused and Geddes refused to go to the United States. There it had become a smokestack industry that was being supported by heavy subsidisation. Japan was a fascinating place, which I visited. At Chiba, there was a wholly new dry dock. There were hostels for the workers in a kind of company town. It seemed to a picture of modernity.

The argument might come in Britain, "but how can we cope with that because the yards are all on Greenfield sites?" But, if you went to another yard in the downtown area of Tokyo itself, next door to a steelworks you would see some pretty antediluvian equipment. But, as Geddes would say you had to look behind the surface appearance. The real advantage was not just that their technology, in places, was excellent, but also the strength of the network, the interconnectedness. Moving to Germany, I saw only two yards. The one I saw in Hamburg was terribly like John Brown at Clydebank. In general, in the countries we visited, there was a lot of things wrong with our industry in comparison, but it was not as bad as all that. That was the defence for this brave new world of Geddes.

Post Geddes; that is a period I find very difficult to be really sure about. My feeling about is that the SIB did not have the sense of fighting the government. The legislation had been passed [*after considerable delay*]. We were now down to the nitty gritty with the government. They were now getting worried about the repercussions of keeping the shipyard lads happy. The SIB did not want to take the considerable difficulty of getting, say, Scott Lithgow on the Lower Clyde into bed together. There were real problems here as well as the question of personalities. They did not want to fight on that and fight the government. Unless the Labour Ministers had been very firm in talking about looking to the future, and stressing the hard road, then there would be difficulties.

The problem was that further assistance was given to the industry, and not in the terms I had advocated. If Geddes was to be implemented in spirit and with the question of confidence, let the government decide what sum should be injected with conditions. It was a grant. The moment you start investing in the industry out come the employers and the unions begging bowls; you have created an atmosphere which is exactly the contrary to the one that you wanted to create.

You were on the path to nationalisation, not only in the technical sense that you actually bought shares, but this was the way it was going. This, I think, is the turning point, and like all real turning points, you can not put your finger on it and describe it exactly. There was a drift. In the end, it is votes that matter to any sensible politician. There was no SIB to stand against it. It became a body which entered into the government. Bill Swallow [*the SIB Chairman*] was a nice man, but not very effective. This was not his world.

UCS had the biggest begging bowl in the country. To put it crudely, I think UCS should have ceased to exist. Whether or not there was a possibility of combining the Lower Clyde with Yarrow, I am not sure.

Geddes arose out of the white-hot technology idea [*used in the 1964 General Election campaign by the Labour Opposition Leader, Harold Wilson, who became Prime Minister*]. It might have come off. It was a terrific gamble. He [*Geddes*] got the people concerned on his side, and the Report got the government on his side. Politics went along with it. I am not being party political in the sense of being against Labour because some of the brightest chaps talking at the time in terms of entrepreneurship were in their ranks. The problem was that the government became a soft touch.

Interview No. 41: Cliff Baylis, Under Secretary (Shipbuilding) Ministry of Technology, Director, Shipbuilders and Repairers National Association

I took up my duties at the Ministry of Technology on the very day the Geddes Report was published [March 1966]. I was in charge of the shipbuilding desk for three years, during which time we were trying to implement the policy recommendations of the Geddes Report. Then I moved on to other work in the Civil Service. In 1973, when I was coming up to retirement, until Tom McIvor and Ross Belch approached me and asked me to take up the job of Director of the Shipbuilders and Repairers National Association, in order, really, to lead the defence against nationalisation. I was happy to do this, holding the post for three years until the SRNA was dissolved. I then did a number of part-time jobs in retirement, including international work on behalf of British Shipbuilders for around five years, and was also a non-executive director of Harland and Wolff at their invitation.

I suppose I can summarise the general government attitude to the industry by saying that it was grossly inefficient and non-competitive in world markets. By the time I was involved it was already in steady decline and was down to ten percent of world output or less. It was generally viewed as pretty old-fashioned and inefficient.

Management was weak, it was inbred. In relation to management, it had one distinguishing characteristic. Shipbuilders quite genuinely believed in what they were doing. They were fanatical about being shipbuilders. They really thought that they were different from others. They really believed that they were the best shipbuilders in the world. To a certain extent, that made them more likeable than many industrialists I had to deal with. It meant, secondly, that it made them much more difficult to convince that what they were doing was not necessarily the best for their future.

The Geddes Report was the bible. At the time I thought it was a good report. I thought it held out good prospects for reforming the industry, although we did not achieve them. I got to know a number of shipbuilders and trade unionists. I read a lot, I talked to a lot of people, and I formed my own judgement. My summary of the situation, I think, would have been generally endorsed by the government departments who got involved.

I personally feel that the Geddes Report was not fully carried out due to a number of factors. It was essentially based on the philosophy of change by agreement, and was not written with a political slant. To a certain extent it was not acceptable to Anthony Wedgwood Benn and the socialist government, because it essentially implied that the industry should remain in private hands, and should be given the chance to reorganise itself with the aid of a certain amount of money administered by the SIB.

The money was conditional on achievement. Personally I do not think the SIB members were as good as one might have hoped. The SIB perhaps, did not put enough pressure on the industry as it should have done. The warship firms opposed, from the outset, the indignity and the financial risk of getting tied in with the merchant shipbuilders. The three warship firms [*Vickers, Yarrow and Vosper Thornycroft*] were always a bit of a law unto themselves, inside the association [*SRNA*] as well as in other matters. With hindsight, one can perhaps argue that they were justified in their attitude. The different owners, by and large, did not really accept the concepts put forward by Geddes. Some, particularly the smaller yards, especially on the Clyde, were only too happy to sell out. In my opinion, they were overcompensated. The appointments made to the new combined organisations, for example, on the Upper Clyde, were not of the calibre required to bring the industry into a competitive position.

I do not think now, although I believed it at the time, that full implementation of Geddes could have saved the industry in the sense of making it really competitive. I think it could have made it much more efficient than it did. Ministers were impatient, particularly Tony Benn and Jeremy Bray, and obviously, even then, were talking about the need to bring the industry under nationalised control, but to be fair to them, they backed the Geddes Report, and they gave us officials and the SIB every opportunity to implement its recommendations. I do not think that you could really criticise the government of the day for failure to carry out the Geddes recommendations. I got on well with Benn, and found that he gave us the full backing that we needed from politicians to try and get it working. I do not think that he was at all surprised when it failed.

I was not directly involved at the time, but I think at the stage of UCS it was a mistake to give way to Jimmy Reid. UCS paid too much for what it took on. Its new management under Hepper were not up to the tasks required. The Tyne was doing rather better with Tom McIvor [*Swan Hunter*] in charge. He was a very shrewd practical and down to earth shipbuilder [*an ex tradesman fitter*] who did not take on more that he thought he could accomplish. They were in better shape than the Upper Clyde, which was really in a mess, and so for that matter, was the Lower Clyde, whose losses was not publicly disclosed, but were huge.

At the SRNA I produced a paper setting out an alternative to nationalisation in which the shipbuilders committed themselves to a greater degree of co-ordination, amalgamation and changes in working practices. Whether they would have done it, had we been successful, is an open question. It was a case for rationalisation rather than nationalisation. On the three-year run up to nationalisation, Conservative Opposition MPs told us that we supplied the best defensive material that they had ever had from an industry. It was partly because of that that the Bill was delayed in the House for such a long period. I have to admit that, although we fought the Bill clause by clause, the shipbuilders' main concern was compensation, particularly the warship firms. Then the appointment of Graham Day as Chief Executive Designate on the Organising Committee did not please them either, because they did not think much of him, and he did not think much of them either. Teddy Taylor [*an ex-employee of the Clyde Shipbuilders Association*] was our closest spokesman. We paid him a retainer. He was extraordinarily helpful.

Nationalisation, as a concept, was something I opposed personally, as did the Conservative Opposition. My opposition to it was based on the fact that building ships is the least appropriate industry for a nationalised management structure. There was a strong and passionate opposition to nationalisation. Do not forget that the Tory Party in those days were terrified that the Labour Party might nationalise whole sections of British industry. The Bill took two years to get through. It coincided with the beginning of the biggest and longest decline in shipbuilding which historically had ever occurred. I can tell you that some of the shipbuilders, not least Tom McIvor, by the time we were two thirds the way through, were saying privately, "for god's sake, let it go through." The situation was getting worse. We were certain that we were going to lose the battle over compensation terms. Some shipbuilders, at least, now could see that delay was doing more harm than good. Within the SRNA, McIvor was in the minority, so we fought it right to the very end. That meant that the Organising Committee really had nothing to do for a long period. Graham Day resigned. I think it would be true to say the delay was damaging to the industry. The key factor was that the world slump in shipbuilding had begun. Nobody, though, could see, at that time, that the slump would be as long and as deep as it turned out to be.

The Swedish industry [*so often held up as a model to Britain*] went. The Norwegian industry almost went, and even the Germans contracted to an extent that was astonishing. Under nationalisation I was aware of the problems facing the industry in relation to the world market, which for the whole period was so depressed that British Shipbuilders never had a chance. I do not believe they could have done it. But given the state of the market there was no way in which any new management could have made the industry competitive under those conditions. Nationalisation, on balance, probably did more harm than good. It perhaps built up hopes that did not exist. I personally think that the style of management and the calibre of the people were not up to the enormous tasks that were facing them.

Interview No. 42: Barry Barker, Director, Shipbuilding Industry Board

I came in by an unlikely route. I had been in India having left for there on leaving Oxford. I was the Secretary of the Bombay Chamber of Commerce, and Secretary of the Metal Box Company in India. We were thinking of returning, when I saw the job advertised with the SIB. It is an important point that the SIB had been set up before I joined it. The secretary [*Pat Hypher*] was a Civil Servant and had been responsible for setting it up I knew because he had been the Trade Commissioner in Bombay. When the SIB was statutorily wound down my involvement with the industry ceased.

The man who did a lot of the leg work was Tony Hepper, particularly around the Upper Clyde. By the time I came onto the scene, the Civil Service had realised that the expectations that they had were not going to be realised. They understood that some chief executive was needed to make things happen rather than relying on shipbuilders to queue up and make submissions to the SIB. There was the usual sudden panic. We moved in about three to four weeks. It was the SIB which finally made the appointment.

One point needs to be stressed; there was no one on the SIB who knew anything about shipbuilding. It was not as daft as it sounded, because they could not have anybody from any of the yards, which were directly involved. I came in fully armed with the Geddes Report and that was about all. By the time I arrived they were about to sign the agreement on UCS. Tony Hepper had made all the running and he left the SIB just as I came in, to chair UCS. Right to the end, I think that the Shipbuilding Industry Act was actually rather good. It was good and it worked, because when the government was funding out grants and loans, it could only be done on the recommendations of the SIB. The Minister could only say yes or no. Of course we would liaise closely, but it did keep us apart from and at arms length from the department, which was very helpful. Given the nature of the Act, we were in a very strong position.

The Geddes Report was the bible, but it was not the only thing because we were in a position to commission studies and undertake market research. What Geddes had to say was not all that contentious, it was certainly very much better about the past than it was about the future. The most fundamental problem to my mind was the industry's failure to embrace modern shipbuilding techniques and the management that went with them.

I was absolutely convinced of the industry's potential for success in the sense that it had the water, the steel and some incredible technical abilities. It was world class in terms of design, and indeed, construction in particular areas. I could see no reason except them, why it should not succeed. Before I came I asked for a tour of all the shipyards to be arranged for me. The civil servants also produced briefs for me which was pretty near useless. They were totally out of date, and were culled from bits and pieces out of files, which had been cobbled together over the years. I got the impression that nobody had ever been near a shipyard. I found that when I went round the industry that the usual pointless things happened like going for lunches. I realised that these were not going to teach me anything. The great advantage of having a SIB of this kind was that I could get up to things that the civil servants could not. I went to the shipbuilders' dinners and parties, and really got involved with them personally. I could get to know anything, and do anything. It was a very happy affair. You got to know that they were a tough lot. The main thing was that there was no reason that the industry should not have been successful as it stood then.

There were two huge problems. One was that they had no clear idea of the market. I do not suppose anybody had. And when we went out and tried to get it, we got it all wrong, particularly on the big tankers. The other problem was technology. The world had learned that in order to control costs that you had to build as large a part of the ship as possible indoors. You were not subject to the weather, and people walking off the job. They had not grasped this, and I do not think it was sufficiently grasped in the Geddes Report. That was the key. If you had done your marketing properly you would have known what market you were in. None of them knew what market they were in. They would do anything. They would build an Austin 7 one year and a Rolls Royce the next. They would go for a standard series of bulk carriers, and then they would want naval work. That was pretty crazy. They were never likely to succeed if they tried to do everything for everybody, and quarrel with the chap across the river at the same time. There had to be new concepts of the market. Companies had to grasp which market they were in, whether the facilities were adequate, and then whether the labour relations were right. I think that the concept of some of the groups was probably misconceived. In the eyes of the politicians it came to be a matter of simply putting the things together. It was not, really, as that was not going to achieve anything by itself.

The SIB was given quite a lot of money. I think it was £50 million in grants, and much more than that in loans. It was pioneering. We were in a fascinating new position. We were not really a banker, but we were the last resort. There was no way that the shipbuilders were going to raise money from anybody, anywhere. Indeed, they could not come to us unless that was the position.

The other thing that has to be said is that in those days you could not move without the trade unions. My chairman [*William Swallow*] was absolutely masterful at dealing with the trade unions. One of the reasons why he was appointed was that he had a great reputation for getting on with the unions. He and Joe Gormley [*National Union of Mineworkers*], who was the vice chairman, were absolutely brilliant. Gormley was pilloried for being an employer's man. Everywhere we went we would see the unions separately. I would never see the management without seeing the unions, it was absolutely necessary.

Shipbuilding was always a bit of a political football. I remember the representatives of Swan Hunter coming to see me on a Friday afternoon. They would tell me that unless there was £x million available for them by Monday there would be twenty thousand people out of work, and what was I going to do about it. They were used to using this as a form of blackmail to get money out of the government. The naval business was where this process was developed to the highest degree of expertise. They were not only expert at screwing money out of the government on social grounds, but also on the grounds of changing designs and tenders and so on. Have you looked at the way a ship was ordered by the Navy? I have never seen anything so complex and bureaucratic. At the end of the day you could make money at every turn of the bureaucratic wheel. It was essentially cost plus business. If you could not make money doing that then you could not make money doing anything.

Was the shipbuilding industry special? You bet it was. It was absolutely singled out. It had its own piece of legislation, its own board, and its own monies, whereas everyone else in the white hot technological revolution had to go to the Industrial Reorganisation Corporation, who dealt with everyone else. At best, shipbuilding was singled out because there was no reason why it should not succeed. At worse, it was because of unemployment, and the social and political problems in all these difficult areas.

The Ministry of Technology officials were superb as civil servants. I got a wonderful view of the Civil Service as it was. I admired Wedgwood Benn a great deal because he never did the dirty on the SIB, and he could have done so easily. The UCS situations as they developed were quite awful, and Benn could have said many times that the SIB had got it wrong. When there was real trouble in the yard with the men threatening revolt; Benn always took on the problem and went to the Upper Clyde. I admired him for that. Otherwise they did not interfere a great deal because I did not let them.

The industry was, despite all the dealings it had with government over the years, quite incapable of providing a real case for money to be given. If we had passed on to the Treasury the rubbish we got from the industry, it would not have stood a chance of getting a penny. Much of my work was really translating the industry's needs into reports.

We reached a stage in the third year *[1970]* when we had got such trust and goodwill out of the industry, and such belief from civil servants and Ministers that I can be absolutely sure that had we gone on the form in which nationalisation would have taken place would have been laid out. I was very clear what was going to happen. The SIB was going to be changed. It would have been a holding company owned by the government. In fact we were postulating, not nationalisation, but national ownership, which was really rather different. It was not going to be owned and run by ministers. We had built up such a relationship with the shipbuilders based on the fact that we had prevented the ministers from interfering. Both sides would have allowed us to have shares. All the people running it would have been from the private sector. The equity, the share ownership, would have passed from one to the other. I was a great advocate of the government possessing shareholders powers and nothing more. You will not find this in the public records, but it was definitely suggested, particularly from Swan Hunter and Scott Lithgow.

On the election of the Conservative Government in 1970, there was a very serious question as to whether the SIB would be allowed to carry on. There was an option under the Act for the SIB to carry on for another year. I think that they thought that the mess would be so awful if they did not let it go on for another year. It was made pretty clear to the SIB that we had a year to clear it up and get out. All that we had built up by way of trust of the shipbuilders who were willing to hand things over was seriously threatened, as was the trust built up in the previous government and Civil Service. The form of organisation I described in the above depended on the existence of this mutual trust. It could have evolved. We could have maintained our position as ultimate bankers. The SIB undoubtedly would have changed with people of real shipbuilding experience coming on. Frankly, it was made clear to us that none of this was going to happen, as there was a very right wing direction coming from the new government. The market was right. Theirs was a form of highly advanced Thatcherism really. If the industry dies then so be it. The truth is that pretty near nothing would have survived, in my opinion. I really do think that the SIB prolonged the evil day for a whole lot of them at vast expense. If we have a shipbuilding industry at all today, it is only because of what was done during that period. After major surgery we had brought the industry into a state of convalescence. I had always thought that a form of nationalisation was always inevitable.

Interview No. 43: Vivian Marchant, Head of Shipbuilding Division, Department of Industry

I became an Administrative Principal in the Civil Service in 1950. I did have a technical and engineering background which was then, and still is, unusual for administrative civil servants. When the post of Head of the Shipbuilding Division fell vacant at the beginning of 1971, I was put into it. I obviously knew nothing in detail about the shipbuilding industry, but within a few months I was confronted with the collapse of UCS, which overwhelmed me and to some extent overwhelmed the government. I held the position for nearly five years until December 1975 when I retired.

When I took over at the beginning of 1971, the situation that I encountered was that there were three companies or groups of companies, which had already had substantial help. One was Harland and Wolff, which always had special importance because they were the largest employer in Northern Ireland. This and the general problem level of unemployment in Northern Ireland worried the government enormously [*and that substantial numbers of British troops had already been committed from 1969 to hold back the tide of sectarian violence*]. Therefore, Harland and Wolff were given special government financial assistance [*funding to cover losses of £10 million at October 1970 and the rescheduling of a £3.5 million loan*], with the government holding a substantial shareholding [*47.6 percent costing £4 million*]. Subsequently, in 1975, the government acquired the rest of the shares, and a government director on the Board. I was very much involved with the Northern Ireland Ministry of Commerce in looking after the interest of the Government as a shareholder.

Things were rather similar, but not so important politically, in the case of Cammell Laird [*at Birkenhead*] where they were again reconstructed [*just before the General Election of 1970, the Cammell Laird Group of companies had announced that it was ten days away from liquidation*] with government help. Cammell Laird's engineering interests were separated, and the shipbuilding side was kept going by a substantial amount of Government equity capital [*through the Industrial Reorganisation Corporation*] and loans, and a government director on the Board.

The third was UCS, with yet again, a government director on the Board. The SIB was still extant. If it had any money left after what it did for UCS, and I am not sure that they did, it was supposed to help the whole of the industry. This was one of the things which caused a great deal of irritation amongst some of the other firms who felt that there were government goodies being handed and that they were not getting a fair share. People on the Lower Clyde complained bitterly, as did others like Sir John Hunter [*the Chairman of Swan Hunter*]. It seemed to them that the unsuccessful were being propped up by government money while the comparatively successful did not get any of these benefits [*up to 1970, Scott's and Lithgows had got £20,000 each in consultants fees to enable them to finally merge; UCS, in comparison, had received some £20 million from government*].

Then the real disaster took place when UCS collapsed. The government's interest was very much that of wishing to maintain employment in these difficult areas. This was the Conservative Government, in spite of the lame ducks policy. I remember Nicholas Ridley's memorandum on the need to butcher the industry. There was remarkably very little difference between Labour and the Conservatives up to this point, because Benn, of course, at the time of the previous difficulties at UCS had made a speech about "so far and not a penny more." This was the same sort of thing that John Davies had said in his speech about no lame ducks. The trouble with both governments from a civil servant's point of view was that no one believed a word that they said. Over and over again the government said that they were not going to give any more help, and then they gave in when the crunch came. They gave in over Rolls Royce, and they gave in over UCS.

It was not the Work-in that precipitated the climb down over UCS. I agree with Frank Broadway's book [*Upper Clyde Shipbuilders: A Study in Government Intervention (London, 1976)*] that the Work-in was largely a fiction. It never really took place [*in the sense of the workers being entirely responsible for the product and being legally liable for it*], except as a publicity stunt. There was a long history, going back to the 1930s of what used to be called regional industrial policy [*the Distressed Areas Acts*]. Under this policy successive governments of both parties accepted a degree of responsibility for the level of unemployment in what were then called the Development Areas. When UCS collapsed there was great concern; not only about the direct increase in unemployment, but also about the indirect effect on employment among their suppliers, many of whom, were also located in the Scottish Development Area. Intense pressure was put on the government in Parliament, and by local authorities, industrial organisations, trade unions and many others. In spite of the brave words about not a penny more, and no help to lame ducks, the government felt unable to resist this pressure. Anxiety about civil unrest was one aspect of the broader considerations, and the threat of imminent revolution on Clydeside was taken quite seriously, they were worried.

In trying to achieve something from UCS, John Davies was responsible for trying to produce some solution to keep at least part of the shipbuilding going, but also to try and find some other form of activity for the yards which were hopelessly inefficient. The John Brown yard was the most serious problem. It had a very long record of losing money in spite of the marvellous reputation it had for the ships it built. It was one of the terrible tragedies that no one could see a future for it in shipbuilding. The trade unions helped to find an alternative for John Brown. A man, James Darragh, from my staff went across to Houston, Texas with a union delegation, including Danny McGarvey of the Boilermakers. They did a marvellous job in getting Marathon to come over and build oil rigs.

The other three yards were amalgamated, on the advice that the problem from the outset must be treated as a social one, to form Govan Shipbuilders. The government agreed to put in £35 million as a sort of first step, a large part of which was for capital investment [*£17 million was to cover losses*]. James Darragh and I invented the name of Govan Shipbuilders although it did not require much invention. We then had to find some directors for it. John Davies recruited Lord Strathalmond, who was then one of the managing directors of British Petroleum to become Chairman. We hijacked a man from Vickers, Archie Gilchrist who became Managing Director. The Finance Director was kept on from UCS, but left soon afterwards.

The next disaster to hit the industry was the Court Line collapse [*in 1974*]. They were a curious company [*over one hundred companies*] partly travel agents [*Clarkson Holidays*], partly shipbuilders [*also a shipping line with eight tankers, an airline and a leisure company owning hotels in the Caribbean*] in the form of Appledore [*a small shipyard in North Devon, also Doxford and Sunderland Shipbuilders, and North East Coast Shiprepairers, Ltd.*]. The company was put into liquidation, but the government did not have anything to do with the tour operating side, but they did decide that something had to be done to help the shipbuilding side of the business, and they were turned into a largely government-controlled company. I was, in fact, a government director for just a few months because of the government equity capital that had been put in.

Ministers [*Labour government of 1974*] accepted that nationalisation was not going to happen overnight. There would be an interim period, and during that period it was necessary, if there was to be something left to be nationalised, to keep the industry going. Some of these shipyards were in such a state that they could not be abandoned for a period if only for the reason that there was a lot of government money already in them. Beforehand, I achieved something that I was very pleased with. I managed to present Harland and Wolff as a free gift to the Government of Northern Ireland. It was an appalling incubus on the Department of Trade and Industry. The situation was so exceptional and highly political there. The amount of money that was flowing into the company on a continuous basis, every year, was quite beyond all reason. It could not be justified on any sort of industrial grounds. It was being justified on political grounds. I was very glad that we in the Department of Trade and Industry no longer had to justify to the Treasury the provision of these tremendous funds, and that we did not have to take responsibility for ensuring that they were properly spent. It was all handed over to the authorities in Northern Ireland and therefore, Harland and Wolff did not come into the reckoning when we talked about nationalisation.

On nationalisation, the ideas which I and my staff worked on were very much the same as those which were eventually put into effect. Broadly speaking a government holding company was to be set up retaining the existence and the identity of the firms. It was a point that was made to us by a lot of people in the industry. It always seemed to make good sense that these yards had names which were worth quite a lot in the sense of long standing reputations.

My view at the time was that nationalisation was irrelevant, and that it would not make much difference one way or the other. I was very depressed about the state of the industry. It seemed to me that the industry was beyond help. It had got to the state where it was hardly possible for any practical policy to revive it. It had gone so far behind, notably the Japanese, but also the Germans and others that it was going to be almost impossible for us to catch up again.

What was remarkable was that when I went to Japan was meeting the workers. Their standard of education and training was in a different world. That kind of thing made me feel that whatever you did in the way of organising, like nationalisation, it was not going to make a lot of difference.

Interview No. 44: Ron Dearing, Regional Director, Department of Trade and Industry for the North of England, Deputy Secretary, Department of Trade and Industry, London

I was Regional Director from 1972 until 1974, based in Newcastle, where shipbuilding was one of the major industries, but had no particular responsibility for it. My second involvement was as a Deputy Secretary in the DTI between 1975 and 1980.

I became involved as Deputy Secretary just when the nationalisation Bill for aircraft and shipbuilding [*the two were never separated*] was under attack for hybridity in Parliament. I was there throughout the following period involved in the return of the Bill after its defeat. There were problems with the Examiners of Bills [*on various challenges on hybridity over the inclusion of ship repair firms*] and eventually the compromise that enabled nationalisation to take place. If I may make one comment, my recollection is that the Conservative Party Opposition at one late stage was prepared to reach an accommodation with the Labour Government if they would exempt ship repair, if not most of the significant ship repairers were knocking on my door immediately afterwards to be brought in. Perhaps the Conservative Party had made a misjudgement. What they might have done instead was to concentrate on the compensation terms of nationalisation which proved very difficult. When the chips were down, and because Labour were not prepared to compromise on ship repair, the Bill fell, and it had to be reintroduced.

The Organising Committee under Admiral Griffin was the shadow Board that was brought in immediately after the Second Reading of the Bill. They had the task of planning for the new industry, and I used to see them regularly. Then there was the division in the Department, the under Secretary, Mike Casey, originally, then Arthur Russell. Shipbuilding had been a major issue in the DTI for years, and it was an extremely tough division to run. The whole history of the industry, apart from the warship firms, had been one of desperate trouble for it to compete.

The obvious difficulty was that the industry was not competitive, and it was impossible for it to get orders without public subsidy, and not small subsidy either. The second difficulty was persuading the Commission of the European Economic Community to allow us to subsidise to the extent that we wanted to. The third factor, at that particular time, was a recession in shipbuilding orders, which had existed for some years. In fact, during my five years it seemed to be one continuing lack of orders.

We had the emergence of the Japanese, and then of the Korean shipbuilding industries, which were immensely competitive. I went on visits to see those industries, and I remember how impressed I was in seeing the [new] Hyundai yard in Korea, where people worked sixty hours per week for the equivalent of about £20. The yard was immaculate; totally clean, no metal lying about. They did not appear to have industrial disputes. The scale of it was awesome. We went to the Mitsubishi yard in Japan where the scale was great, and the organisation of work was highly sophisticated.

Coming home, it was not the case that there had not been investment in our yards. For example, the Pallion yard [*Doxford and Sunderland Shipbuilders on the Wear*] was excellent for the type of ships it built. There had been heavy government support in the form of investment, not always, but there had been a great deal.

When the Government refused to compromise so forcing a reintroduction of the Bill, Graham Day, the Chief Executive Designate on the Organising Committee resigned, because in his judgement it was too late to save the industry. He felt that an opportunity had been missed, there should have been a compromise, and was very critical of the Government. He was invited to a private interview with the Prime Minister, James Callaghan about his resignation, but did not recant. Once Day left, there was no Executive really left. Mike Casey was then drafted in from the Department to do Day's job, and others from the industry were brought in. It was difficult for a very good man like Mike Casey, who was a civil servant, to go in and take over an industry in crisis, and replace someone like Day who was such a successful experienced manager. This implies no criticism of Mike who was a hero to take the job. It was just that he had not the experience of Day, who is one of those people who had outstanding managerial ability and experience. He was a major loss to the industry.

The industry was structured in such a way that there was a holding company in effect. Each of the shipyards was property in its own right. Under Graham Day there would have been strong leadership because he takes tough decisions, clearly and quickly. We got a good Board together though. It included people from the unions, Danny McGarvey. John Parker, Bill Richardson and Ross Belch came on the Board, so we got the strongest people from the industry.

One aspect in which I was deeply involved, which showed both the Government's commitment to the industry, and the tragedy of it all was the Polish order. It was Graham Day who spotted the opportunity and opened negotiations. The Polish prime Minister came across shortly after Day had resigned, and the two Prime Minister's agreed that the order would come to Britain. There were twenty-eight ships. This was the one big order in the whole world. It was a great coup. Unfortunately, and understandably, the price negotiations were not for the Prime Ministers, it was left to British Shipbuilders and me. The prime ministerial agreement had been made public. I was involved in these negotiations and eventually the order was delivered. The strategy of British Shipbuilders was to use this order to save the Tyne and Wear. British Shipbuilders were also determined to use the order as a means of getting improved work practices, which were essential if the industry were to survive. British Shipbuilders wanted the end of demarcation. Danny McGarvey was immensely powerful with the Boilermakers on the Tyne. He and Ken Griffin [*another union Board Member*] led the pitch to the workers, but they were rolled over. The workers would not have it. British Shipbuilders, therefore, decided that they would not place the order on the Tyne. They went to Govan Shipbuilders which was saved [*and which was already marked for closure beforehand despite huge government investment*]. However, the ships were not suitable for Govan, as I remember it. The Tyne was the right place for them. I have forgotten what the losses were on the Polish order, but they were huge, perhaps £100m including the subsidy. I had to go before a Parliamentary Select Committee to render account, which was a memorable experience. It was a tragedy to the industry because immense effort had been devoted to getting the orders, and beating off the world wide competition. It was a tremendous victory to get the orders. I am not blaming anyone, but it was a tragedy that agreement could not be reached. But, British Shipbuilders felt that that was the kind of change that was needed. Even with the knowledge that thousands of jobs would go, it was not a price that the workers were prepared to pay. I think it was one of the biggest setbacks. There were thus two tragedies; the loss of the leadership that had been preparing for a year, and the loss of the order, which could have recreated the heartland of the industry. Thereafter, we struggled to get orders.

The process of getting orders was collaborative, and the structure was very simple. British Shipbuilders designed the ships and pitched for them. Our role in the Department of Trade and Industry was a modest one. It was to respond to bids from British Shipbuilders for grants, which would enable them to reduce the price that they quoted. It was a subsidy. The whole of the European shipbuilding industry faced the same problem. All countries were subsidising. The European Commission though that it was madness for countries to be competing in this way. They issued a control of the subsidy with rules. It was the job of the Department to consider proposals form British Shipbuilders for subsidy. Then the job was to fight for them in Brussels, having secured the consent of the [*British*] Treasury, which was never straightforward. It was a very tough job. In Brussels they had a section, which was very strongly led and, as you may know, the traditions on the continent are different from here. They tended to be legalistic. Considerations which are outside the rules are irrelevant. I went across to argue with the Minister, Gerald Kaufman, but it was pointless. You could put forward an argument like this, "but we need this to provide a framework with which the management can do a deal with the workforce to improve industrial relations to get major changes, etc.," and you would be told that it was not relevant.

There may have been strategic reasons too. There was even a Scottish dimension. I was concerned to get orders for Scottish yards, because I realised things were stressed [*rising tide of Scottish nationalism*]. But these broader issues were outside the EEC Intervention Fund rules. This was immensely frustrating and absorbed immense amounts of time. It slowed down the ability to win orders. I expect other countries felt just as we felt. The role of the Department, therefore, was to evaluate the requests from British Shipbuilders for support. We had then to secure the Intervention Fund, and each individual application had to be forwarded separately.

You have no idea what it was like in Europe to get things through, and this was on top of the British Government's system; the Treasury, the Department needing to be satisfied that this was public money well spent, and which could be justified to Parliament. There is all the difference in the world between a successful industry and one which is in dire crisis, and totally dependent on the Government, and through the Intervention Fund, the European Commission, for orders.

I am totally non-political but I have a lifetime's experience of public ownership. One of its tragedies is that it was always identified with the centralisation of ownership and control. What I have found as a manager is that the more you de-centralised, the more commitment, the more enthusiasm and the more progress you make. There does have to be a framework within which people want to move of their own volition in a particular way. I regard centralisation as one of the fallacies that has been built into the thinking of the Labour Party on the question of public ownership. In the case of British Shipbuilders, the industry had been in trouble for years and years, and every time there was a downturn in the world market the industry needed government assistance.

Interviews
The Politicians

Interview No. 45: Tony Benn, MP, Minister of Technology, Secretary of State for Industry, Secretary of State for Energy

I was made Minister of Technology in the summer of 1966. I had a historic interest in the shipbuilding industry because my mother was a Paisley woman, and I lived in Clydebank for a while as a child. From the election defeat of 1970 to 1974, I was shadow spokesman for industry, and in 1974 I was made Secretary of State for Industry, before being sacked by the PM Harold Wilson in 1975. I then had ministerial responsibility as Secretary for Energy from 1975 to 1979, which gave me some sort of involvement as in shipbuilding as the Marathon yard at Clydebank built oil rigs.

In the summer of 1966, Harold Wilson called me and said that I would be in charge of a very big department. One of my first jobs was to implement the Geddes Report. We had the SIB, whose job was to try and deal with the decline of the industry by setting up viable estuarial units. My first task as a new minister was to get these going. I got into a lot of trouble with George Brown [*Deputy Leader of the Labour Party, First Secretary of State and Minister of Economic Affairs at Department of Economic Affairs, and Foreign Secretary later in 1966*] who had set up the Fairfield Experiment, because he thought that UCS would undermine Fairfield.

Any Minister will be told, in effect, to rely upon the Civil Servants, who actually do not know much about it. They are all generalists who move about perhaps doing shipbuilding one week and the aircraft industry the next. The Civil Service provides the briefings. Then, at the Ministry of Technology, I had an Advisory Committee on Technology, which included scientists and industrialists. I also saw the shipbuilders and the unions. I always made it a key issue to keep in touch with the shipbuilding unions, of which there were a large number. One of the problems of the shipbuilding industry as compared to mining was that there was a mass of unions. Inter union rivalry was quite difficult and was used as an excuse by those who were hostile to the industry. I had a fair range of advice open to me, and I always tried to keep sources of advice that were not official.

I started with a little knowledge of the shipbuilding industry, but I did pick it up. You had to assess who would trust for information. I made the union involvement direct to me and not through the Civil Service because it was very unhelpful about the trade union movement. They did not really like the trade unions, to be honest. They did not know what to do with them. I remember once at the Department of Industry talking to quite a senior official. I suggested that he contacted the Confed [*Confederation of Shipbuilding and Engineering Unions*] and he said, "what's that?" He had never heard of it, and he was the senior civil servant on industrial policy. It turned out that Whitehall regarded the unions as people who were dealt with by the Ministry of Labour or Employment. As far as industrial policy was concerned the unions had no function. I think it is an important point to make because it explains a lot.

I got more information through my intimate relationship with the unions that any other Minister I knew. It meant that instead of being a constitutional monarch who said yes or no to what the Civil Service proposed, I became an umpire between the competing interests. That meant that I could take a much more independent, and hence, political view. The impression that I had of the old shipbuilding families was that they were Victorian in their character. If they were not making a profit, they could not afford to invest. They had no marketing. They would drop into the best clubs and meet a ship owner, and they would say "I need another tanker." They allowed their workers to work in appalling conditions. I went to Glasgow in January 1969, and went to a yard where I addressed the workforce on a freezing cold day. I told the men that I was having difficulty getting money, and with absenteeism at eighteen percent, I could not do it. One man shouted out to me "if you had to work in this weather cutting cold metal in the open air you would be absent eighteen percent of the time." Of course, he was right. By then the Swedish and Japanese yards were cutting steel in proper covered places. These guys were literally out in the icy cold cutting metal. I am only giving you anecdotes, but I think they quite influenced my mind.

The owners were the old families. The industry was run by what you might call the hard hat people. They had come up from the bottom. They had no training in management, but they knew the men and respected them. They were replaced by a third group of people who were the MBAs from Harvard who had their sleek grey suits and spent all their time on cash flow, and would turn up at the yard at 11 a.m. in a Jaguar car, and would float in and out. The industry for them was another way of making money. [*Benn is talking about consultants here, as Harvard MBA holders and the like do not accord with the managerial structure of the industry*].

I do not want to insult Tony Hepper, but he came in from Pretty Polly Stockings. He was put on the SIB. Originally, I think he had been one of George Brown's industrial advisors. He was then made Chairman of UCS. He was nice enough, but he did not know anything about the industry. His remedy was close, close, close unless you could be subsidised, subsidised, subsidised.

It was at that point that I first came to the conclusion that if you did invest in the industry, you did not just invest in the old management, you had to try and bring the workers in and make a joint effort. That was the beginning of the industrial democracy idea that the guys had an interest in the future of the industry and should be equal partners. That was absolutely revolutionary to the owners and the hard hats. I have a sneaking admiration for the hard hats, because they did know the industry. They were harsh and brutal, but the men had a respect for them. They did not have any respect for the old owners, and they did not have any respect for the business managers. I found myself more at home with that crowd. I had a lot of time for Ken Douglas who respected and recognised the skill of the industry.

The other problem was the [*Upper*] Clyde was too narrow to launch big ships. There was thus a geographical problem which could only be resolved by building new shipyards on Greenfield sites. The money was never there for that. On UCS I managed to get the money to keep them going, but the Cabinet said that this would be the last tranche. This was used continually by John Davies, who would say, "Mr. Benn said that this would be the last tranche." I could not get the Cabinet to give me any more money.

We were picking up an industry which was in the advance stages of deterioration. I think one thing that George Brown did at Fairfield was give people back some hope. On UCS I tried to do the same by stressing that there was a future for the industry. We had to take some decisions and they had to be jointly agreed. I got shouted at by the workers because there were sometimes lay-offs. I did it myself. I did not go through the various boards. I tried to get the SIB to help me as far as they could. The people I found most difficult to deal with were the old owners.

My honest conviction was that the British establishment was so angry at the power of the trade union movement that they were prepared to sacrifice British industry to break the power of the trade unions. It was a sort of deliberate policy of demoralisation; nothing we do is any good. The difference between that and the atmosphere that can be seen in the wartime films with cheerful women making Spitfires is striking. There is a strong psychological element in all this. I was aware of it coming from the Civil Servants, and from the *Financial Times* and the BBC. On Civil Servants, I do not want to sound paranoid, but these men lived in the Home Counties [*around London*] and travelled up on the train every morning, or, if they were very senior, in their chauffeur-driven cars. They knew nothing about British shipbuilding. This is not a Scottish Nationalist point, but it could be made as one. London was not remotely interested in Scotland, or for that matter, Bristol, Wales, or wherever. These were the natives and they were causing trouble and would have to be disciplined. I would actually have mass meeting in my own department, and sat there with all these civil servants who had never seen a Minister and told them what we were trying to do. Some were quite sympathetic, but they would be surrounded by the old Praetorian Guard of the old Oxbridge network. The main planning document in Scotland was the railway timetable south to London. If there was a crisis they would get on a train and come to London and plead with Ministers. I hated St Andrews House [*the home of the Scottish Office in Edinburgh before power was devolved to a Scottish Parliament in the 1990s*]. Willie Ross [*Secretary of State for Scotland*] was no help at all. There was no financial discretion whatsoever in Scotland.

There were real problems with shipbuilding in terms of policy, even if I had got all that I wanted to get the show on the road with a real investment programme, it would have been very difficult. We did not own the industry. Bullying and bribing [*in the sense of investing in yards*] businessmen to get them to do what you wanted them to do was really like pushing on one end of a piece of string. It was that period, 1966-1970, in terms of my own political development, that I became a Socialist. I had always been a Labour man. Seeing the difficulty of operating by bribing and bullying, and boards and intervention when it was so obvious that there was an industry which needed to be re-equipped, and set up with a fresh start, to have got rid of the old owners, to have encouraged the hard hats and let the industry be run by the people who knew how to run it. That really shaped my thinking. The Cabinet took the view that we should do no more than administer the status quo more efficiently. One reason that I could have not joined the Social Democratic Party [*a breakaway group led by four former Labour Ministers who formed their own party, the SDP, in the 1970s*] was that I was already a member of an SDP cabinet from 1966 to 1970. We were picking winners and using all the sort of language that has now reappeared in the Commission at Brussels. It was all about incentive and penalties. It was awful; and of course, the industry died under our feet while we were at it.

The radicalism that appeared between 1970 and 1974, encouraged by Ted Heath [*the Conservative Prime Minister*] and his lame ducks, enabled the real policy to emerge. Those were the seeds and the fruit grew in opposition. We then tried to harness it in 1974, but that was too much for Harold Wilson. That is the story of 1966 to 1970, going in full of keenness. It did not work. We then worked out a proper policy, discussing it totally with the unions, and then coming back [*to government*] in 1974 [*twice*] and introducing the nationalisation of the shipbuilding and aircraft industries etc. Wilson did not really believe in common ownership [*nationalization*] and was bitterly critical of workers control and industrial democracy. Being an old Civil Servant himself, the unions were people you sort of kept in with. You used them at the annual Labour Party Conference to discipline the left. They paid the party cash, but when you got into power they were the trouble-shooters for you if there was industrial trouble. This was his attitude to the unions. My attitude was wholly different, and had been shaped by my experience of trying to run the Ministry of Technology policies.

I was never a great believer in the Morrisonian [*Herbert Morrison, Labour Party Deputy Prime Minister, 1945-1951*] model of nationalisation. Everything that I nationalised [*not the shipbuilding industry*] I kept much closer; there would be the power of the specific directive. Otherwise you were just observing a nationalised industry. Coal and Steel were totally outside any ministerial control. I think in the case of the shipbuilding and aircraft industries there was a power of specific directive so you could call in the management and unions. It was tripartite in character. The old owners did not want nationalisation. It was a case of shifting the balance of power. I am the author of the phrase, "a fundamental and irreversible shift in the balance of wealth and power in favour of working people and their families." I never saw nationalisation in terms of cold competition. It was the recovery of self confidence of a people and their capacity to produce with the proper and reasonable support from government; and the fact that the old capitalist class no longer ran everybody like serfs - that was the political inspiration for it. I had at the same time to say to them that we had to earn a living. I think they understood that. You had to be practical about it, but the object was not just to produce better figures in the third quarter.

On the ideological opposition to it, ideology suggests philosophical. It was class conflict really. There was a school of thought that the working class had become so strong that it was running the country. It never got anywhere near running the country. The cutting back of the perceived power was what is was all about. It was done by employment legislation, by media campaigns against working people, and the trade union movement and so on. They saw me as a traitor to two classes; my social class, and the governing class. [*Benn, then Anthony Wedgwood Benn, had renounced a seat in the House of Lords*].

Privatisation was the final blow. If you threw them out in the cold to survive, and to compete without support, it was not privatisation, but a death sentence, and I think almost consciously so. One of the factors that must have encouraged the Government was that the oil [*North Sea*] was pumping ashore. Industry was seen as quite unnecessary. That is when the moment of truth will occur, when the oil runs out, and we find that we do not have a strong enough industrial base to earn a living as a nation.

Looking back at the whole period since 1945, I think that the truth is that the Labour Party stopped doing, not technological indoctrination, but explanational activity. They left the field totally free to the ideologues of the right, who were clever at exploiting discontent and attacking inefficiencies. We never spoke about the power of capital, and what it was doing to us. In 1979 when we lost, it was not so much that we had been defeated, but that we had surrendered years before.

Nationalisation in shipbuilding should have happened after the war, when the mining industry [and steel] was nationalised. Whatever is said of the nationalisation of the mining industry, it did lead to better health and safety. There would have been a huge investment, and we might have taken off at that stage in a way that would have provided some serious opportunities for the industry to recover. At the end of the war we had it easy with much of the competition being destroyed. There was a huge demand for our goods. We never used that period properly. It would have been a big change.

Interview No. 46: Norman Tebbit, MP, Secretary of State for Trade and Industry

My involvement with shipbuilding began during the nationalisation process, and my next acquaintance with the industry occurred in 1979, when as Parliamentary under Secretary of State I was responsible for merchant shipbuilding. For a brief spell I was number two to the Industry Minister, Keith Joseph, with responsibility for shipbuilding among other matters until October 1981. I then left to become Secretary of State for Employment and returned to shipbuilding again between 1983 and 1985 as Secretary of State for Trade and Industry.

I was struck by the extremely poor industrial relations, the widespread restrictive practices, and the short-term thinking on both sides of the management and workforce and unions. I had very little sympathy for some of the management who struck me as continuing to behave as if though they were the proprietors of great shipyards in the nineteenth century when we dominated the shipbuilding world. In fact, they were essentially state employees, and should have regarded themselves as being alongside their workforce. They still sat in their oak-panelled boardrooms with the pictures of ships that had been built in the last century. They still enjoyed their lunches in the traditional style. They, then, somehow expected the workforce would not react adversely. I think there was extraordinary managerial incompetence. It was scarcely surprising that the unions were as badly led as they were. That, of course, did not apply to all the yards. There were some yards where there was really quite a high level of investment, where labour relations were much better, and where management was vastly superior. A number of those were naval yards. Even there our lack of competitiveness was growing as can be seen by our progressive failure to sell warships in the export markets. There were some which were not bad, and by British standards, pretty good.

In general, the weaknesses outweighed the strengths of the industry. International competition was unfair, but our industry was in such a parlous state that it was not subsidy by overseas builders that was the problem. The problem was the sheer efficiency of the yards in Japan, and more recently, Korea. As against European yards there was a very clear problem of protectionism, most notably by the French and Germans, who would not allow orders to go to British yards, or indeed to other nations. This was even when our yards were capable of meeting them on level terms.

In my years of government, so far as the European aspects are concerned, we were particularly interested in reducing or eliminating subsidy, and in lowering the capacity of shipbuilding yards in Europe to something more in line with what the market required. We aimed to ban unfair subsidy and covert protectionism. We achieved rather more on paper than we did in practice. The French are willing to sign anything; but whether they are willing to stick to it is distinctly another matter. We had the problem of the Greek yards as well where similar considerations applied. Having said all of that, the real problem was the super-competitiveness of the Japanese. Tragically, for example, we produced a superb yard at Harland and Wolff in Belfast for building very large ships just at the time when they were going out of fashion.

I had a great deal of sympathy for the workforce. After all you could not expect them to have a much greater and wider vision than that of their managements. They were poorly led. They were insecure, and the saw the solution to their problem as protection, subsidy and restrictive practices. Now, to a large extent, that had worked for a number of years so you could hardly blame them for thinking that it would continue to work. We went down the wrong path as we did in the motor industry of refusing to accept that when markets shrink then so must the production facility. It was better to actually close yards than to reduce the scale of production in all the yards in trying to keep them open.

I do not think nationalisation was necessary. I think it was undesirable. I think it delayed the potential solutions, and therefore some yards which might have survived have perished. I do not think that anything could have saved most of those yards before the twin centrepieces of Thatcher policy came along. One was the reform of trades' union law. Alongside it was clear policy of the Government which came to be understood that if managements of workforces destroyed their businesses, they would face the consequences for themselves, and would not be bailed out.

I was not exceptionally impressed with most of the management of British Shipbuilders. There were too many of them emotionally attached to the idea of building ships as opposed to the reality that shipbuilding was not an activity in its own right. It had to be an activity which was generated in response to a demand from customers for the product. I think it was a tragedy that Graham Day did not come into the industry as it had originally been intended. Had he got to grips with the industry earlier I think that things would have been better. However, I do not think that even he could have done a great deal to save the industry without the other changes which came more generally after 1979.

Post-1979 I concluded that we would have to come back to saving what yards we could, in particular the naval yards where we could see, roughly, the volume of business which they would be likely to achieve. It was unlikely that the Royal Navy building programme would go overseas. Clearly we could take out those yards and they would have a reasonable prospect of survival subject to the volume of Royal Navy building. Then we had to look at the merchant yards to see which of them could survive. There was also the question of the engine makers, which was another story. Here there was the enormous capacity to be measured against the likely demand for their product. We just had to get down to reducing the size of the industry to doing what we could to create a stable regime in co-operation with Brussels, in which potential investors could see what was ahead.

Interview No. 47: Teddy Taylor MP, Scottish Office Minister for Industry

After I left university I got a job with the *Glasgow Herald*. Seeing an advertisement for a job as an official with the Clyde Shipbuilders Association, I applied and was successful. The CSA primarily handled labour negotiations, and made representations to the Shipbuilding Employers Federation and to Government. I was with the CSA for five years, during which time my job was basically yard negotiation. I then represented a Glasgow constituency where a lot of people worked in the shipyards, and became for a brief period a Scottish Office Minister taking the industrial brief just before the General Election of 1974. I then became a shadow cabinet spokesman, and was prominent in the debates leading to nationalisation.

People like us in the industry did not look upon the Geddes proposals as anything that was going to solve problems. We felt that the problem had come. There was nothing in setting up things like the SIB. It gave you a new avenue to government whereby you could argue for more shipbuilding credits. My own view was that the Geddes proposals were basically irrelevant. There was almost inevitability with the problem at this stage. All we did with these brand new plans was to create a few jobs for a few smart people for a few of these business consultants.

To be brutally frank, if I had been in charge of government and the shipbuilding industry, I do not think that there was anything I could have done. I got the impression that the battle was lost before it was started. I do not say this as a criticism of the shipbuilding industry. It is the sort of thing that is partly inevitable when you have got an established industry and when you are dealing with unfair competition, and, more importantly, a declining demand for ships allied to a huge increase in capacity. If anybody believes that simply investment is the answer, then they should visit Harland and Wolff.

Governments were all terribly anxious to do something about shipbuilding, because they genuinely believed that something could be done. They had innumerable committees and experts looking at the industry suggesting various things. I hope that I am not sounding pessimistic, but there was little they could do. I think we might have achieved more if we had sent out an aeroplane dropping pound notes at random over the shipbuilding industry.

I had a view that that there was just a possibility that UCS, if they had established links with other countries, might have won a lot of orders. These would have been left wing countries because the UCS people were left wing. There was just the possibility also that they would somehow motivate their people to perhaps work in the way that the Japanese did. I think that the only prospect was that as a collective enterprise it was possible to achieve new standards of achievement out of people which otherwise would not have happened. This did not materialise.

I think that the period of the 1980s has been acutely depressing because I just can not see how it is possible for a UK shipyard to compete with a Japanese shipyard when there is a declining world market. It just can not be done unless you were to start all over again with completely new practices, organisation and image.

I think that we could have achieved more if we not gone for structural change of setting up the SIB and nationalisation. We would have done more by doing nothing in terms of setting up a national organisation on a national basis. I think money was wasted on unnecessary re-organisations. There would have been a sort of survival of the fittest. Moreover, firms would have realised that there was not some huge national organisation which might have helped them out of their problems. Firms should have felt more desperate than they did do.

Interview No. 48: Bob Clay, Labour MP, Sunderland North

I became involved with the shipbuilding industry when I became the MP for Sunderland North in 1983. At that time there were a number of shipyards in my constituency, which at a later period became re-grouped under the heading of North East Shipbuilders Limited. Sunderland was still one of the principal UK merchant shipbuilding centres and one of the main sources of employment in my constituency. I have simply been a constituency MP who has acted as a spokesman for the industry and the workforce during the period of its rundown in Sunderland, and nationally.

The period in which I was involved was when the Conservative Government privatised the warship yards and some mixed yards. It was a period in which many people argued that this would be the beginning of the end, that you could not separate the warship yards from the merchant yards. That was proved a correct analysis. With hindsight, the real cutting edge of that argument was not so much that it was totally wrong thing to do, but that corporate separation might have had logic to it provided that there was a clear policy for merchant shipbuilding. The problem was that there was not. So the industry internationally was going through an appalling period, and the UK merchant yards just closed year in year out, one after the other.

With the advent of Lord Young as Secretary of State and Kenneth Clarke as Minister of State at the Department of Trade and Industry it is quite clear that a decision was made in late 1987 to write of merchant shipbuilding. Shipbuilding was in such a state and the market was in such a state that it could not conceivably be profitable. I think that they then went through this guise of saying that if people want to buy it they can. If they do not want to buy it, it will close, knowing perfectly well that most of it would close. Govan got the Chinese ships deal, which made the yard suitable for privatisation. Then the Government got through the Kvaerner deal [*a Norwegian shipbuilding conglomerate which at that point could not access European Intervention Fund monies, but could after its purchase of Govan*] to buy Govan.

The basic purposes of the European directives were purportedly to slim down the European shipbuilding industry to a competitive state, and to basically allow the subsidies set by directive in exchange for a reduction in capacity. Graham Day carried out what his Government told him to do [*he was required to do this in any event*]. His successor, John Lister, made clear his dissatisfaction at losing Govan Shipbuilders. [*Thereafter, having lost the warship sector, and the two large composite yards, Cammell Laird and Swan Hunter, British Shipbuilders was left with only one medium yard, NESL, and two small yards, Appledore and Ferguson, which were earmarked for privatization. Therefore, no critical mass existed in the merchant sector to enable British Shipbuilders to survive, and the end of volume shipbuilding in the UK was already apparent*].

I would say that the Conservative Government saw shipbuilding, more than any other, as a dirty, clapped out sunset industry.

Interviews
British Shipbuilders Plc

The portents leading to the nationalization of the British shipbuilding industry in July 1977 through British Shipbuilders Plc were not propitious. Despite the growing market challenges facing the industry, the period from 1974 to 1977 was characterized by internecine political bickering over the nationalization of the British aircraft and shipbuilding industries in Parliament. The debate soon descended into an ideological dogfight that unduly delayed the process of nationalization under two Labour governments. It also had a serious impact on the shipbuilding industry's ability to survive against withering international competition. On the industrial front, the road to nationalization was not aided by the shipbuilding firms which through their trade association, the Shipbuilders and Repairers National Association – formed in 1967 to replace the old Shipbuilding Employers Federation, Shipbuilding Conference, Dry Dock Owners and Repairers Central Council, and National Association of Marine Engineers – virulently opposed state control. Conversely, the shipyard trade unions whose members' jobs were at stake were almost messianic in their desire for nationalization. The delay resulted in firms paralyzing development or investment plans and maintaining very high levels of manning.

By the time nationalization actually took place, however, most firms in the industry were unprofitable and faced an uphill fight to survive. Owing to the long delay in nationalizing the industry, the Organising Committee Chief Executive Designate, the Canadian lawyer and shipbuilding general manager (Cammell Laird), Graham Day, resigned, and Mike Casey, a career civil servant, took his place. British Shipbuilders' first Chairman was Admiral Sir Anthony Griffin, a former Controller of the Navy and Third Sea Lord. Griffin resigned in 1981. Prior to nationalization, the ship repair industry was excluded by an amendment to the original Bill when it was before Parliament. Only after nationalization were six additional ship repair companies acquired by British Shipbuilders Plc. Post-nationalization, the shipyards, marine engine builders and ship repair firms taken under state control at 31 March 1978, were Appledore Shipbuilders Ltd.; Austin and Pickersgill, Sunderland; Barclay Curle Co. Ltd.; Whiteinch, Brooke Marine Ltd., Lowestoft; William Doxford and Sons, Sunderland; Cammell Laird, Birkenhead; George Clark and North Eastern Marine Ltd., Sunderland; Clelands Shipbuilding Co. Ltd., Wallsend; Goole Shipbuilding and Repairing Co. Ltd., Goole; Govan Shipbuilders, Govan; R, and H. Green and Silley Weir Ltd., London; Hall, Russell and Company, Aberdeen; Hawthorn Leslie (Engineers) Ltd., Newcastle; John G.

Kincaid Ltd., Greenock; London Graving Dock Co. Ltd., London; River Thames Shiprepairers, London; Robb Caledon Shipbuilders, Leith and Dundee; Scott Lithgow Ltd., Greenock and Port Glasgow; Scott Lithgow Dry Docks Ltd., Greenock; Silley Cox Ltd., Falmouth; Smiths Dock Co., Ltd., Middlesbrough; Sunderland Shipbuilding and Engineering Ltd., Sunderland; Swan Hunter Shipbuilders Ltd., Wallsend, Walker-on-Tyne, Hebburn, South Shields and Haverton Hill on Tees; Swan Hunter Training and Safety Company Ltd., Wallsend; Tyne Ship Repair Group Ltd., Newcastle; Vickers Shipbuilding Group Ltd., Barrow-in-Furness; Vosper Shiprepairers Ltd., Vosper Thornycroft (UK) Ltd., Woolston and Portsmouth; Yarrow (Shipbuilders) Ltd., Scotstoun; and Yarrow (Training), Ltd., Scotstoun.

As laid down by the Aircraft and Shipbuilding Industries Act, 1977, British Shipbuilders Plc was to promote the efficient and economical design, development, production, sale, repair and maintenance of ships and slow-speed diesel marine engines and research into matters relating thereto, and, to have full regard in carrying out its activities to the requirements of national defence and to promote industrial democracy in a strong and organic form in its undertakings. Initially, British Shipbuilders headquarters were in Knightsbridge, London, but were swiftly moved to Benton House, Newcastle-upon-Tyne. British Shipbuilders had a difficult beginning (in its first nine months of operation from July 1977 it posted pre-tax losses of £108 million) and an even more difficult future. Mercantile contracts were aided by Intervention Fund assistance under the Industry Act, 1972, but under the Treaty of Rome no such assistance was given for warship contracts. Nor did the Intervention Fund apply to ship repairing. British Shipbuilders, up against withering international competition, soon began to rationalize its assets by closing shipyards and restructuring its divisions. Redundancy schemes were instituted, and negotiations with trade unions centralized. Perhaps its most disastrous decision was to enter the large offshore structures market under then-Chairman Robert Atkinson by creating an offshore division with Cammell Laird and Scott Lithgow as the main yards. This resulted in horrendous losses. The Conservative government under Margaret Thatcher decided to privatize British Shipbuilders from at least 1983. Under the chairmanship of Graham Day, who had agreed to return, Scott Lithgow was the first constituent firm to be privatized in 1984, sold to the industrial conglomerate, Trafalgar House Plc, for a knockdown price of £12 million, with British Shipbuilders picking up the bill for redundancy payments. The mainly profitable warship-building firms soon followed, leaving British Shipbuilders with an unprofitable rump of merchant shipbuilders, ship repairers and marine engine-building firms, some of which were later sold and others closed permanently. By 1985, the Thatcher government had pumped some £1.4 billion into British Shipbuilders. Employment, which had stood at 38,000 in 1977, had slumped to just 5000 a decade later. When North Eastern Shipbuilders Ltd [an amalgamation of Sunderland Shipbuilders at Pallion and Aus-

tin and Pickersgill as Southwick] was closed in 1989 as part of an alleged underhand deal by the Thatcher government and the European Commission to reduce shipbuilding capacity, British Shipbuilders was left bereft of any major shipbuilders, having sold Govan Shipbuilders to the Norwegian conglomerate, Kvaerner, in 1988. The remaining engine-building firms were closed or sold, and British Shipbuilders remained only in name for legal and insurance purposes until 2011 when it was officially dissolved.

Interview No. 49: Admiral Tony Griffin, Chairman, British Shipbuilders Plc, 1977-1980

I joined the Navy as a cadet in 1934, and spent forty-two years there. The last five years I spent as Controller of the Navy, and it was there that I became closely involved with the shipbuilding industry, especially the warship builders. I left the Navy in December 1975. I think it was largely because of that experience and because various trade union leaders had declined the job of Chairman of British Shipbuilders that I was invited to do it. On the morning I left the Navy, I had my first meeting of the Organising Committee the same afternoon. I was only too glad to do so because I had seen enough of the shipbuilding industry to know that, especially as shipbuilders were severely threatened, something needed to be done urgently to co-ordinate them, and bring their very substantial skills to bear on a highly hostile world situation.

Having collected a team largely selected by Graham Day [*Chief Executive Designate*] we set about our job with a view to becoming a formally established corporation in the middle of 1976. We had to work extremely swiftly to consult with the industry and the unions, and to work out our plan. Our broad objective was, first, to co-ordinate the very substantial talents in the industry. The warship yards were among the best in the world with very advanced facilities. Secondly, we had to eliminate several weaknesses. To these ends we needed money to promote and sustain operations, and the authority to apply new ideas. This, essentially, is what nationalisation did, and meant so far as I was concerned.

It did not mean the same thing to a lot of other people, notably the workforce who felt that nationalisation was a way of simply securing their jobs whether the company was solvent or not. So a lot of our time was spent bringing home the reality of the international scene, where it was clear that in the post-war period the developing world in many areas adopted shipbuilding as an industrial plank in their economic recovery and development. They were perfectly prepared to engage in major loss leading activities to this end. This crippled not only British Shipbuilders, but practically all Western European yards, including some of the best in the world, like the Swedish shipbuilding industry. The competition was massively subsidised, and had few problems with industrial relations because the South Koreans, for example, were prepared to settle one or two strikes with armoured cars.

However, the Organising Committee were able to make progress, reckoning that necessary changes could not be brought about sensibly without the maximum involvement of all concerned. That did not just mean sitting down and talking to trade union leaders. We were very sensitive about the need to consult middle managers too, and a team went round every single yard at which we invariably had three meetings: one with the Board, one with middle management and one with the trade unions. Before going on these visits, yards were sent questionnaires outlining our thoughts of where the industry should go, and inviting a response before we went. We were able in this way to discuss their views. There were very positive responses to the whole idea of involvement. But, on the national political scene, the [*Labour*] government had a majority of one in the House of Commons, and a minority in the House of Lords. Whereas some people were advocating nationalisation for doctrinaire reasons, the Tory Party was opposing it for opposite doctrinaire reasons. Nothing that anyone could say would change these two positions at a political level. Everything was done to frustrate the Bill. The issue of hybridity was used to frustrate the Bill.

As a result we had long passed our intended date of vesting in July 1976, and saw no sign of the Bill being enacted. At that stage, in November 1976, Graham Day lost his patience and resigned and went back to Canada. His example was followed by three key directors for finance, personnel and planning, and there was no longer any Organising Committee except myself. I think this spurred everyone, not only the government, to break the deadlock. It was done by the Government deleting a handful of companies from those listed as ship repairers in the Bill. This was accepted by the Tory Party who too realised they were placing a major industry in even more serious trouble.

Once that was accepted, early in 1977, I was given the go ahead, and asked to replace what had been lost. During that time the only former Director, Ken Griffin, who had been on the point of leaving like the others, but changed his mind and stayed on. Between us, with the Government's help, we reconstructed bringing in Mike Casey from the Department of Industry, where he had been running the Shipbuilding Policy Division, as Chief Executive. We were then able to collect a new team together, and get our proposals approved by the government, and that led to vesting in July 1977.

It was in the course of this that we established unique relations with the trade unions. It was very fashionable at the time to go in for industrial democracy, and worker directors, and so on. From what we saw of this in action in other nationalised industries it really did not seem very productive because the so-called worker directors were elected by their people, and this meant that they were mandated to them, and could not really operate freely without reference back. Secondly, as soon as they became Board members, they lost their trade union emotional connections, and were regarded by many of their former colleagues as one "of them," and were soon distrusted, and discredited, and became of limited use.

By sheer chance, we ourselves were very lucky because through the contacts we had made around the industry, two or three top trade unionists were constantly dropping in to the office, and were always made welcome. We would always have a chat, and they would occasionally produce useful papers. So when the time came to establish their support more formally, they were simply invited to join the Board without losing their trade union positions and without being elected. We were able to get them formally appointed by the Secretary of State, and so they had full status, but at the same time retained full connection with their own people. It was an extremely effective channel of communication between both sides, and it was lucky that one of the three was a boilermaker, John Chalmers [*hull trades*] another was an electrician, Ken Gregory [*outfit trades*], and the third was from the General and Municipal Workers Union, Ken Baker [*labourers*]. They were all worth their weight in gold.

Once British Shipbuilders was formally established we had the authority to organise and address the world wide shipbuilding problem, and set about increasing efficiency and competitiveness. There were many aspects to this. Not necessarily in order of importance, we were able to establish a centralised marketing operation for the entire industry, so that customers knew exactly who to deal with, [Individual yards could still chase orders however] and we were able to trade from a far wider base than any individual yard had been able to do in the past. Industrial relations were obviously an extremely important feature, largely because people came to understand the international marketing problem, they accepted that changes had to be made. We were able to persuade the government that these changes in the shape of reductions of people could not be made solely on the terms of the standard redundancy scheme available in industry. We managed to introduce a special redundancy scheme. Through this, and other means, we were able to achieve major reductions, all redundancies being voluntary. We needed to close yards, not only because that was dictated by the market, but also because the EEC, who had a finger in all this, refused to allow any kind of subsidy to the industry unless it was for restructuring, that is, for yards that were to be reduced in capacity or closed. Many yards were closed and at the end of three years – in 1979 – something like 24,000 people had been taken out of the merchant shipbuilding sector. At the end of that time, the time lost on industrial disputes was the lowest on record. People had become aware that the object of nationalisation was not to preserve jobs at any price but to ensure that the country had a lean and fit shipbuilding industry that could survive and compete world wide.

We spent a lot of time amongst British ship owners inviting them to place, if not all their orders, then at least some with British Shipbuilders so that we could re-organise ourselves and ensure that their source of supply was not entirely in the hands of foreigners. In some cases, a minority said that they would never build abroad, and they did not. They were first class. Others recalled the scene after the end of World War II when ship owners urgently needed to replace ships that had been sunk. The shipbuilders of the day were in a sellers market, and in many cases dealt somewhat cavalierly with their customers. At least one distinguished owner of a major shipping line told me that the boot was now on the other foot as he could get all the ships he wanted more cheaply from abroad, and without our industrial relations problems. I do not blame the ship owners as they had an increasingly difficult time trading, and they did not want to spend a penny more on their ships than they needed to. The national interest of keeping a source of alternative supply cut very little ice.

Almost without exception, the shipbuilders were opposed to nationalisation from the outset. They felt that some extraneous authority was going to tell them what to do, change their names etc. However, as Vesting Day approached, a number of firms who were in severe financial difficulty, and one of which was trading illegally, could not wait for nationalisation to happen. After the Board had made and maintained contact with them, and spelt out what it was trying to do and why, things changed. They could keep their names, staff, have local negotiating rights, and they would then be responsible once the marketing had led the customer into their yard. Once they understood these ideas, I do not think a single one objected. The warship yards were left fairly well alone. Their fundamental objection was the same as the rest, but it tended to endure longer because they were not in serious trouble. They had full order books and a captive customer in the shape of the Royal Navy, so they were not as anxious as the other yards.

International competition was unfair, and the most conspicuous example was South Korea. The prices covered no more that the cost of raw materials, which were practically the same universally, and all the rest was subsidised by the South Korean Government. This had previously applied, to an extent, to Japan. The main advantage enjoyed by the Japanese, was that the whole enterprise was regarded nationally. The whole enterprise was addressed in terms of "Japan Incorporated." This meant that the government departments, the finance houses, the banks, equipment and materials suppliers, shipbuilders and also ship owners all contributed their bit to being able to undercut world prices. It was, also, to some extent, subsidised, but it was the co-ordinated national effort which really counted, and shipbuilders knew that they had the wholehearted backing of their domestic owners.

There were others who were almost as bad. Brazil had a major subsidy programme, the Poles subsidised in a major way and so did the Chinese. It was all the same game which pretty much destroyed the West European shipbuilding industry. It was a fundamental problem, but it was not the only thing to blame. A lot of the problems were self inflicted.

Our government was a great help in getting the redundancy scheme through, and in getting EEC approval for the Intervention Fund, which was their response to grossly unfair competition. It undoubtedly helped us to keep afloat during the early years. However, the government's subsequent involvement was counter productive. We found that because we were a nationalised industry the department concerned insisted on knowing exactly what we were doing and why so that they could justify themselves in detail to the Public Accounts Committee, and the Treasury. This led to deeper and deeper involvement in our affairs until a stage was reached when every single contract had to be approved by the Department of Industry. That was not the end of it. The Department of Industry found itself beholden to the Treasury for the Intervention element of each [*merchant ship*] contract. At the same time, as this Fund was regulated by the EEC, a department in Brussels headed by Viscount D'Avignon, also insisted on vetting every single contract, and this caused further delay amounting on average to four months. All this deterred customers to such an extent that I personally asked the Chancellor of the Exchequer, Denis Healey to help. He declined to do so.

The officials concerned were apprehensive about their responsibilities as accounting officers. They had to be prepared to answer detailed questions about the industry of which they had little prior knowledge or experience. Everything had to be explained and justified to them and they were not prepared to delegate to the people they paid to run the business. My successor, Robert Atkinson, saw the government heading for the fragmented privatisation of the warship yards, and he personally had to see eight Conservative Cabinet Ministers to get that stopped at the time. This privatisation was eventually undertaken for entirely doctrinaire and non-commercial reasons. Altogether, dealing with government was a constant struggle, which deflected high grade people from their proper task of competing internationally. There were one or two exceptions amongst Civil Servants who realised what an appalling pitch they were giving all our nationalised industry Chairmen.

There was a radical difference between Labour and Tories. Labour really did want the industry to succeed, and saw the significance of British shipbuilding to an industrialised island country. I had very little difficulty with Ministers on that. The Tories regarded the whole thing as a lame duck, costing a lot of money, and as far as they were concerned the sooner it was gone the better. This was not immediately apparent in the change of Government in 1979. I was reappointed by the Sir Keith Joseph, the Conservative Secretary of State. I think that the Tories were uncertain as to our chances of survival. They were under EEC pressure to decrease the Intervention Fund, which was gradually reduced, but at too fast a rate. It ceased to be effective, and the government eventually settled on a figure which bore no relation to the competition. As a result we could not win enough orders, and the industry was eventually closed down. The Conservatives showed no conception to what this meant for an industrialised island country.

On bureaucracy, I do not think that any sort of organisation is without criticism, and I accept that. Headquarters staff did become much bigger than it should have been, and if I had a free hand I would have tried to keep to a promise I made very early on which was based on my talks with Arnold Weinstock. His GEC ran a similar sized operation, with 120 companies from a London office with a staff of eighty-two, and I reckoned I could do the same, even although we had to have centralised marketing. I think that there was inadequate control of our management organisation. Nevertheless, I think that the way British Shipbuilders handled nationalisation, including decentralisation of operations to individual yards, all of whom kept their original names, created a reasonable chance for us to have held our own. We could be in that position now if we had been allowed to follow our own course of action.

Our main constraint was financial authority. We could not negotiate our contracts freely, even although we fully accepted that intervention funding was not unlimited. We kept pointing out the tortuous process of taking seventeen steps to get authority. The time this took destroyed the whole objective because the customer was not prepared to wait. This meant that we were hamstrung. We volunteered and were prepared to guarantee, that if we were given the authority we would not exceed realistic guidelines. This was not accepted and customers just went elsewhere.

The warship yards were a very important and very highly priced sector of The British shipbuilding industry. I think that there was a failure on the part of us all to go sufficiently for exports. The last export order for a major war vessel was for Brazil in 1970, long before nationalisation. The order was still running when nationalisation started. Throughout my time we were unable to win a single export order for a major war vessel. This was highly regrettable, because it meant that the warship yards were left in a monopoly position with just one customer, the Royal Navy. Their fortunes depended entirely on how the Navy's vote went. They were not cushioned at all by export orders, even although we spent much time in Algeria, Argentina, Egypt and China trying to get export orders simply to even out the work. One of our problems was that the Royal Navy's requirements were too sophisticated and expensive to be within the purse of our former customers. Foreign requirements were being met, for example, by the Italians, who, with government support, marketed a special kind of export frigate. We could not get the Ministry of Defence to lower its sights. I do not blame them entirely, because they had a high quality threat to deal with, and could not really afford to come down. There was not enough support from the government for our exports. So we are now in the position of having only three specialist warship yards, Vickers, Yarrow and Vosper Thornycroft, with Swan Hunter as a mixed warship and merchant yard. All have difficulties because they have become too beholden to the Royal Navy. Much of this is due to lack of co-ordinated appreciation of the market, and insufficient concentration, in this case, on exports, and the significance of exports to this country.

Interview No. 50: Michael Casey, Chief Executive, British Shipbuilders Plc, 1977-1980

In 1975 I was appointed Head of the Shipbuilding Policy Division at the Department of Industry. I was responsible for advice to Ministers on policy, and detailed monitoring of specific shipyards which had come within the ownership and control of the British Government. Government were by this stage responsible for Cammell Laird, Govan Shipbuilders and the Court Line companies such as Doxford and Sunderland Shipbuilders and Appledore. I was appointed at the end of 1976 to replace Graham Day as Chief Executive of what was then called the Organising Committee of British Shipbuilders, and in July 1977 I became Chief Executive and deputy Chairman.

The shipbuilding industry's labour relations were extremely poor and its management was extremely poor. Almost every aspect of the industry's structure and functions needed to be reviewed and dealt with. There was a need to drive into the industry at every level a whole range of reforms. That really was the initial task of British Shipbuilders at a very difficult time. The market had changed and British yards had simply failed to recognise the extent of these changes. They had clearly failed to recognise the massive impact of new competition on their share of that market. The new investment proposals which were developing in piecemeal fashion were coming far too late in the day, and were addressing only one particular part of a very comprehensive series of problems.

I have no doubt whatsoever that the great majority of companies outside the defence field were likely to collapse at or around the time of public ownership, or very shortly afterwards, and indeed a very significant number of them were gravely financially embarrassed. I do not think it is true that the shipbuilding industry opposed the Bill in reality. The great majority of the companies desperately wanted to be brought into public ownership, and were deeply concerned at the delays in case they would collapse beforehand. The proof of the matter was that this was a perception overseas, and the Government had to introduce, before public ownership, a Performance Bond Scheme, which it underwrote. This meant that the British Government was providing a performance guarantee with relation to orders placed with the yards in case they collapsed before the ships were delivered. This was very necessary. I have no doubt that the industry was as sick as any industry could be. The only exceptions were some very small yards that were not subject to public ownership, and possibly some of the warship yards.

The delay to the Bill had the effect of delaying or deferring necessary improvements and the industry was in such dire straits that it followed, quite obviously, that the patient did not get to hospital very quickly. I inherited the then Organising Committee, which had become almost defunct with the departure of Graham Day and colleagues. I inherited an outline of the Corporation in terms of its structure but there was no flesh on any of those bones, and nothing by way of a corporate programme. The first task I had was to create an organisation which could form the new Corporation. That meant the recruitment of an entire Board, finding offices to house the Board and the Headquarters personnel. It meant finding a Headquarters for the Corporation which Graham Day had said he would form in Merseyside. Eventually, we decided, on my recommendation, to locate in Benton House in Newcastle. The setting up of the Corporation was a very substantial task. Then there were all the financial, accounting and reporting systems that had to be planned and put in place. All of these were parallel tasks, I hasten to add. Then there were the industry-wide policies of improving productivity and labour relations etc, which had to be devised. We had a statutory duty to produce a Corporate Plan, and to lay it before Parliament. We had international responsibilities and so on.

The initial advantage was that the Corporation was funded to enable the industry to have a breathing space, to be modernised, rationalised and have the financial backing for new investment and marketing. We also proceeded to market the business of British shipbuilding on a world basis, co-ordinating the work at the centre with the work of individual yards. One of the orders obtained was the largest shipbuilding order placed in this country, which was for twenty-four ships for Poland. That was a long and arduous negotiation. That order stabilised many of the shipyards and enabled a number of things to happen. It allowed us to introduce very radical new working practices for massive improvements in productivity, because it provided both work and an opportunity, against that background of success, to persuade the trade unions to agree to completely different working practices. Some of them just would not accept, like Swan Hunter.

The industry in this country had to be rationalised because the industry world-wide had contracted massively. We really had to reduce the number of shipyards on a structured basis. What we were trying to do was to stop a chaotic situation developing, which would have been the case if not for nationalisation. Even with public ownership we had to come to terms with the complete changes that had been taking place over the previous twenty years in the space of about twenty months. The reduction in the size of the industry had to be managed. That was really what it was all about.

We needed to have a hard core of yards that would be capable of meeting international competition. This entailed modernisation, improvements, better management, better practices, better labour relations, and much higher productivity. We produced a Productivity Improvement Handbook, which was practical and laid out in simple language, and we distributed it to every worker in the industry.

In fact, there could not have been the slightest doubt whatsoever about respective responsibilities between the centre and the yards themselves. Our policy was to have the maximum amount of devolution to the yards. There was nothing to stop the yards winning orders themselves. They were all encouraged to do so. We introduced structured systems under which they had to do things which every company in every industry has to do, but which were unknown in the British shipbuilding industry. They were proper financial controls, proper reporting systems so that we knew exactly what they were doing. When we mounted the central marketing operation we needed to be kept informed of what they were doing, which some of the yards seemed to object to? All I can say is that if those shipyards had been subsidiary companies of any organisation I have ever had anything to do with these sorts of complaints would have been regarded as being completely fatuous. The respective responsibilities were extremely clearly set out. The fact is that these companies did not like having to pull their socks up and having to move into better management practices. Some of them moaned, but quite a number did not. The whinging managers knew that their tenure in the industry was probably going to quite short. That particular type of complaint was wholly unfounded.

On the accusation of over-bureaucracy at the centre, every business that is taken over will complain about the burden of reporting, the amount of supervision, the amount of paper. In point of fact, I and my Board went out of our way to be as frugal as we could in terms of what was imposed. If anything we were worthy of criticism for not enforcing more ruthlessly some of the disciplines which needed to be imposed on bankrupt firms. They had no money whatsoever. Their cash flow was being funded by British Shipbuilders. They did not know when they were building a ship whether it would make a loss or a profit! Clearly, there were a thousand and one things in this industry which needed to be redressed. All of that did entail discipline and action from the centre. It is inevitable that this would have caused resentment amongst some people who had to work a bit harder than they had done previously.

Shipbuilding was very backward because of its organisation and the fact that it was in slow decay. It thought that it would never be allowed to collapse by any government, and therefore was quite complacent. You could see this in the attitudes to delivery dates, to sensitivity of clients, to marketing. They had lost markets through their neglect, through arrogance, and through incompetence. They lost the customer base that they thought they would be able to retain through the UK ship owning fraternity. Bleats about British ship owners are a typical sort of excuse, a crutch, for a management which was unable to persuade them to order.

At my level I can say that the Labour Government provided very significant support. The detailed operations of things like the Intervention Fund and the problems of the trade union movement were things that I expected. These were not insuperable obstacles. These were quite justified. Can you imagine the support we got from government from a series of private banks? I can not.

I was with British Shipbuilders until the summer of 1980, by which stage a Conservative Government was in power. The difference between that administration and the previous one was public property. Labour wanted public ownership to work. It was prepared to face major unemployment and redundancies to that end with the objective of a much more efficient industry around a hard core capability. The Conservative Government's programme was to privatise the industry, which meant dismembering British Shipbuilders. There could not be a more fundamental difference than that. What do you do with the concept of selling off shipyards during the greatest slump that has ever taken place? There was no plan for the survival of European shipbuilding as there was for steel, which had much wider powers to introduce quotas etc stemming from the original European Coal and Steel Community.

Given the way in which the pattern of events has unfolded, and the pattern of political change, I do not think there is anything much more that could have been done in that period of time from 1977 to 1980, that would have made any difference. We would still have not had sufficient time to bring about the sort of transformation. It could not have been brought about in the space of eighteen months or two years under one government, and then one year under a government that had a different policy. In fact British Shipbuilders only had a lease of life as British Shipbuilders for two years. If British Shipbuilders had been created a decade earlier it would have been very much better placed to have met the very serious downturn and recession that took place in the second half of the 1970s.

Interview No. 51: Ken Griffin, Deputy Chairman, British Shipbuilders Plc, 1977-1983

In 1970 I was asked as a trade union official on loan to join the Department of Industry as an Industrial Advisor to advise across a whole range of industry from a trade union perspective, and was designated as a Deputy Secretary. At the time the Secretary of State was John Davies. I was the first ever advisor to be loaned from a trade union. During my first year the UCS crisis broke, and in historical terms, plus the collapse of Rolls Royce, that spelled the end of the Heath policy of non intervention. I stayed on after the first March election of 1974, to advise Tony Benn. I also stayed to advise the new Secretary of State,

Eric Varley after Benn had left the department for the Department of Energy. I then helped with the passage of the Nationalisation Bill. Ship repairing, to me, was an irrelevance and I told Varley that. However, it became a political phallic symbol that the Government could not lose. The Speaker's decision was that the Bill was hybrid. I was responsible, among others for recommending that Graham Day should lead the industry. Eric Varley chose Admiral Griffin. After meeting him, I was concerned and said to the Permanent Secretary, "if you appoint him you will have to cover him because he has no experience of real life." As a result they agreed, and asked me to do it. After Day and others had left, only I and Admiral Griffin remained. We had to appoint a Chief Executive at short notice and chose Mike Casey who was extremely erudite and knowledgeable, but had never ran anything. After British Shipbuilders was formed I was appointed Deputy Chairman, and remained there until 1983.

I can confirm that UCS was Tony Benn's road to Damascus. He met real people there. By the time it was finally over, in my opinion it was because of a report from the Chief Constable of Glasgow warning that there was the likelihood of another Belfast in Glasgow. Nationalisation was put in the Labour Party manifesto of 1974 on the insistence of McGarvey and the trade unions. In my opinion Tony Benn was a right winger when he was Minister of Technology. The man who had recruited me was Otto Clark, who was considered one the brightest civil servants in the department. I met him after Benn's return in 1974, and his response was what has happened to my Secretary of State? He had changed tremendously.

The delays to the nationalisation process caused huge damage to the industry. I think it was the difference between succeeding against immense odds and becoming impossible. That was not my view at the time, although it was Graham Day's. We had lost almost twelve months. We had lost a group of people we had chosen to run the industry. We had lost the impetus. We did not cross the line running as we should have, rather we staggered across.

Had nationalisation not occurred, Robb Caledon and the Lower Clyde would have collapsed. Govan would have had problems, and so too would Sunderland. Cammell Laird was bust, whatever government that was asked to oversee that type of collapse would have had very real problems. What we actually did was to phase the industry down with the agreements we had from 85,000 to 45,000 without rancour. We had disputes, but they were very few.

By 1980, the Conservatives had decided to dispense with the services of the Chairman, Chief Executive and the Board Member for Finance. They approached me through a Minister [*of State*] called Adam Butler, and he asked me to stay on as Deputy Chairman. I would have to accept the fact that during my period of stewardship they were going to privatise the industry, and I was not prepared to accept that as a condition. I turned him down. They came back to me with the beautiful prose of the high civil servant, which meant that it was never mentioned. I was the only one of the senior people who was reappointed in 1980. I stayed there for the whole of the next three years until 1983. The period of time I was there I was extremely unhappy for two reasons. First of all, it was perfectly obvious that the Government was running out of patience. In that regard I must tell you that we made substantial progress. We got to within £10 million of breaking even. Ministers are only in office for two years on average and they want results in that time. It was not possible. Secondly, we had a Chairman who was difficult. I was anxious that somebody ought to come in and follow me and Atkinson who would give the industry a chance. I phoned Halifax University [*Nova Scotia*] and spoke to Graham Day, and asked him whether he would be interested in coming back. I then went to see the Permanent Secretary and told him that the next appointment would be crucial. Day took over once Atkinson had received his knighthood.

Atkinson had split the Corporation into four divisions [*actually five*] merchant shipbuilding, warship, oil rigs, and ship repairing [*also engine building*] with a Chief Executive in charge of each division who was responsible for his actions because there was a meeting every month or two months when the man said what his Corporate Plan was, and what were his intentions. In two months time he would have to explain if and why he had not met the targets he had earlier proposed. There was not an arms length relationship. There was a dialogue as to what these problems were. We knew the risks of bureaucracy, but given the scale of the problem we could not fail to apply a set of rules which kept control.

When Day had taken over from Atkinson it was part of his initial plan that I would continue to help him. That was not possible on two counts. Firstly it would not have been fair on the Personnel Director, and secondly, I do not think that you can go back. Day has strengths, but he was a political child. He, unfortunately, did not have anyone who could act as a political adviser to him personally. What he did was to take the ministerial line and put it into operation. The line was to break even in a certain time. That meant savage cuts; not the evolution of change that I was in the business of. We had civilised the trade unions. We had treated them as equals, as they were entitled to be treated. They had responded in kind. Suddenly, after six years of this treatment they were cast out.

Interview No. 52: Michael Haines, Acting Finance Director, British Shipbuilders Plc, 1977

I was a career Chartered Accountant, and in 1975, whilst employed by Thomson McLintock I was seconded at Under Secretary level as Deputy Director of the Industrial Development Unit at the Department of Industry. It was established by the Conservative Government after the passing of the 1972 Industry Act to assist in the restructuring of Industry. When the Labour Party came to power in 1974 it was used more in the way of supporting lame ducks. In shipbuilding terms the lame ducks were principally the Drypool Group based on the Humber, Robb Caledon at Dundee and Leith, and I was also involved in appraising capital investment projects at Cammell Laird and at Sunderland Shipbuilders. Thus in my two years there prior to nationalisation, I became quite heavily involved in the shipbuilding industry and interested in it. I returned to my accountancy firm in January 1977, but was later invited by Mike Casey to become Acting Finance Director from March 1977 in the Organising Committee of British Shipbuilders until such time as they got a full time person in post. That was Maurice Elderfield who took over round about the end of 1977. During my time there I was actively involved in the recruitment of the permanent finance team who actually remained more or less in place until the end of British Shipbuilders.

I was involved in setting up the accounting system, and the rather controversial matter of centralised banking. We agreed that on the first day of nationalisation all the companies would remit their cash into the centre. As firms like Vosper Thornycroft, who had fought nationalisation pretty hard, had £50m in the bank, this was not a popular decision. I was involved in making people understand that they no longer controlled their own destiny, and that investment decisions would be generated and approved centrally. I was also involved in acquiring, on behalf of British Shipbuilders, the six ship repair companies that did not wish to remain independent. We brought in consultants on that in the hope really that they would tell us that we would be daft to do it. They produced a report which enabled us to go ahead. In fact A&P Appledore produced the best mixed metaphor I have ever seen in a report. It said that there was, "a carrot on the horizon."

I could understand the logic behind nationalisation, and was most enthusiastic about it. The traditional civil servant at the time I was in the Department had a hearty contempt for British business, and for the British businessman to the extent of not even understanding that no matter how bad they were at their jobs they were perhaps more relevant to the nation's well-being than being a top civil servant. There was a tendency to fall in love with Charlatans, and get bored by solid, rather boring people who were doing the best job they possibly could. I am bound to say, having worked very hard to try to involve government and industry for the period I was there, that looking back on it, one achieved very little. Unless a business is good enough to stand on its own two feet then it should not be there. It just so happens with shipbuilding that it should either have been a monopoly or it should have been a nationalised industry. To have so many yards competing for everything was a recipe for disaster.

This is a funny story about the British Shipbuilders' $65 million Bond [*Eurodollar Loan*] that was issued in 1977 to finance the Polish deal. We worked very hard with Hambros to borrow the sum at nine percent interest repayable in 1992. It did not have a formal government guarantee. We all reckoned it was an extraordinary good rate [*it was*] and what is more Hambros underwrote it despite the fact that that the underlying business of British Shipbuilders was not at that time profitable. We were very pleased with ourselves. However, at the very last moment, Harold Lever, who was Chancellor of the Duchy of Lancaster, and a great interferer in things like this said that a fixed rate of interest of nine percent was for old ladies, and that we should go for a floating rate. It was agreed that a deputation from British Shipbuilders would have to go and persuade him to the contrary. Because I had dealt with Harold Lever on the Chrysler affair when I was at the Department of Industry, I said, "leave this to me." I know where he lives. Admiral Griffin, Mike Casey and I went to Lever's home in Eaton Square. I pressed the front doorbell and the door was opened by a butler. We sat down and said we were from British Shipbuilders. The butler looked puzzled. The walls were covered with pictures of the Honourable Mrs Vere Harmsworth, which rather surprised us. Then the awful truth dawned on me that I was on the wrong floor. We then beat a retreat! We then persuaded Harold Lever that nine percent was reasonable.

Certainly the best chance the industry would have had was if the whole industry had been flattened about 1950, and specialist yards built at strategic places with concentration on particular ships.

Interview No. 53: Reg Arnell, Board Member for Finance, British Shipbuilders Plc, 1977-1991

My discipline is chartered accountancy. I joined Vickers at Barrow in 1954. After a number of jobs at various Vickers concerns I joined Smith's Dock and after they merged with Swan Hunter I became Company Secretary of Swan Hunter. I then went back to Smith's Dock as Financial Director. On nationalisation I became Financial Controller of British Shipbuilders, then Director of Finance, and subsequently Board Member for Finance. I was made redundant in 1991.

Prior to British Shipbuilders I thought the industry's major weakness was a lack of marketing skills. When I worked at Vickers High Walker yard in the 1950s we built a ship for the Orient Line, and made a genuine, very large profit. The Chairman of the company gave some of this money back to the ship owner against my advice. I thought this was absolutely ridiculous. He disagreed, saying that the reason he was doing it was because the ship owner was a friend of his, and that he wanted to stay friends with him. He said, "when the difficult times come this man will remember that and he will be loyal to us." The very next time Orient asked us to tender, they went to the Far East. We lived in this false world where we thought out friends would be faithful to us. If it had been me I would have gone to the cheapest market as well. In the old days things were incredible; I have seen a contract of sale not signed until after the ship was delivered, because it was all done on trust. Overheads were kept to a minimum, which was one of the reasons we were poor on marketing and on estimating. In the old cost plus days there was no incentive in getting the estimate right.

Even before nationalisation the government were supporting the industry financially, and continued to do so until 1988 when the Conservative Government said, "enough is enough." The reason they stopped supporting us was not because we were making losses, but because we could not predict how much those losses were going to be. I had to go back to government and say month after month, "you know I told you that I wanted £100 million for this year as a subsidy. I am sorry, but it is now £110 million." They would ask me if that was the end, and I would say that it was. Then I would go back the next month and ask for another £5 million. So you could not blame them. It was not so much the extent of the subsidy, although that was an element, but the fact was that it was a progressively deteriorating situation, which was not under control. They were not controlled losses.

We did make a mistake with nationalisation. We could have had some sort of tie up with the ship owners. I am not saying that we should have nationalised British ship owning, but we could have saved both shipping and shipbuilding to the benefit of both. We could have had a comprehensive marine policy. The idea would have been to give British ship owners subsidies, tax advantages and so on providing they did a number of things. One of them was that they had to build in British shipyards. If they had to, you could have said to them give us a forward projection of what ships you need. We could have then loaded Govan, for example, with the ships that they could build best for the next "x" years. If we had organised ourselves with a proper maritime policy through the government that I think that we could have saved some of the merchant fleet and our own indigenous shipbuilding, both of which are now gone. I think that the help went to the wrong place. British shipbuilders should not have been subsidised, but the British ship owners should to have helped them compete with the rest of the world. The future load of work would have been beneficial to the shipyards in a number of ways. If you can not plan your work in a continuous flow, then overheads do not get absorbed. Labour is not fully utilised.

One of the things you could argue against this was that it would have reproduced the culture of the 1950s when there was plenty of work, and everybody said that there was no need to worry since there is a marketing device which ensures continuity of ships. That would have been a danger which we would have had to deal with. It might have encouraged slackness.

A continuous flow of work is necessary. At Govan, for example, there was a period of about six or seven months between the North Sea Ferry and the Polish ships. That was just a dead loss of fixed overheads with no income. That was one of the worst cases. I think that if I remember correctly, that a quarter of the overheads were never recovered. A lot of the cost occurred because you could not load up the shipyards in an efficient way. The gaps would mean that you would get the steelwork finishing on one ship and the steelwork for the next not commencing for several weeks. It was not worth sacking the men, so they hung around doing nothing. A lot of the inefficiency stemmed from that.

On nationalisation I think that we missed a trick. It did not bring any changes to the industry whatsoever. We had access to a bottomless purse, which lulled us back into a false sense of security that our jobs were safe. I am not sure that nationalisation as a general principle is good. I think there is a case for saying that market forces have a place, although, perhaps there should be some controls. I do not think you should take away the market forces away altogether. The threat of bankruptcy is not a bad threat providing it does not become *fait accompli* I do accept that there are advantages with nationalisation. One is the strength that government has to influence other industries, which have an impact on you. Government to government talks can be beneficial as Callaghan did over the Polish deal. That was one of the success stories, although Callaghan was silly to say to the world that we had the contract before the contract was signed. The basic concept was good. It was that sort of thing we should have developed. Margaret Thatcher did it with China, saying that she would give aid if they bought a couple of ships from British yards. I suspect that both Labour and the Conservatives did it for political reasons rather than as a continuous process.

Again, on nationalisation, firms should have been broken down, and not allowed to continue as autonomous units. The industry could not have survived without nationalisation. Maybe the answer was to have nationalised and to have put proper resources in to put it back on its feet, and then to have privatised it again. It should have been privatised in a way that told the industry that it had to stand on its own two feet. We either had to have nationalisation or lose the industry. In the event, we got the worst of both worlds because we both nationalised it and lost the industry.

Interview No. 54: Dr. Martin Stopford, Group Economist, Director of Business Development, British Shipbuilders Plc, 1977-1988

I joined the shipbuilding industry indirectly in 1971 when I was recruited by a firm of London-based shipping consultants, Maritime Transport Research, which had been set up in 1966 as British Shipbuilding Exports and served the SRNA by producing marketing reports. By 1971, I think the management at Maritime Transport Research had found that it was an impossible task to undertake marketing for shipyards where they had no executive authority, and they changed direction into carrying out research studies on bulk cargoes and future markets for the type of ships built in the UK. I remained on the payroll of the SRNA and we were located in their offices in London until nationalisation in 1977, when the SRNA was wound up. On vesting day, I joined British Shipbuilders as Group Economist. In September 1980 I was made Director of Business Development, responsible for corporate planning, and held that position until April 1988 when I moved to Chase Manhattan Bank.

One of the surprising things about the shipbuilding industry was that it had a great deal of knowledge of its problems. Most of the young men who had compiled the Patton Report on Productivity became senior figures in the industry before nationalisation. The conclusion I came to at the end of spending ten years throwing money at the problem of why British Shipbuilders productivity was not very good was as follows. The best place to start is probably with understanding just how bad it was. In terms of comparing productivity you will be told by all the experts that you can not do it precisely, and that is true. But, it is a fact that the Japanese shipyards that were, say, producing Panamax bulk carriers, were taking about 300,000 man hours at a time when British shipyards at nationalisation were taking about 950,000 man hours. In fact by the end of the 1980s they were taking over 1.3m man hours, so productivity actually fell during the ten year period; despite the fact that there had been quite an increase in sub-contracting. Shipbuilding is the world's most difficult industry. I have worked in four industries now and none are as difficult as shipbuilding. It is a very complex industry. It has got a very long gestation period for the product, and it takes years and years to do anything about it.

The British shipbuilding industry never did move from the craft base to the production planned industry, despite a lot of evidence that people knew that that was the way they had to go. The Patton Committee said that. The real power centre was on the shop floor, and any move towards production planning required that the power centre moved towards the planning office, and, of course, they never even got their noses in the door. The planning office became the home of bright young men who were encouraged not to go too much on the shop floor. It was a bit like a research department. It never really got integrated.

Then there was the second element, which is fairly apparent if you go to Japanese shipyards. It is the so-called junior managers, the charge hands and the supervisors and yard managers are all graduates. By contrast, the tradition in the UK was that people were recruited from the shop floor. The trouble is, to make these planning systems work; you have got to be a little bit intellectual. In the 1960s in the UK we had the great build up of university education, which meant that anyone who was bright got siphoned off to university, what you were getting in shipbuilding was really the bottom end of the spectrum. I always felt that you got a culture set around middle management. On the basis of some psychometric tests on middle management undertaken towards the end of British Shipbuilders at NESL, these managers had a distinct personality profile. If I recall it correctly it contained a very strong conservative element. They were people who were very loyal to their bosses, very hard working, not terribly bright, but not dim. There was a character mould there and you can not change a company organism overnight, not when it is the size of a shipyard. We threw money at the problem, and we knew all the answers, and if you got back over all the numerous corporate plans I think we did the right things. We got consultants in, we got advisors. We knew what we ought to be doing with production planning, material control systems, and palettisation units, pre-outfit units, cad-cam-you name it and we tried to do it. We planted the seeds but we did not have the ability to grow those concepts into reality. So instead of accumulating and getting a little better every day we actually got a little bit worse every day. So that in my view is why productivity did not ever go up. It was just that the businesses never got their act together, and accepted that it was worth changing.

In the first couple of years at British Shipbuilders there was an idea that we should focus on certain products and do development work with owners. On a chemical carrier design, for example, we did this in a textbook way. We set up a team to develop a product, which we believed in and we knew there was a demand for. It was a complex product which we believed we could get some technical advantage out of. We put in some technical experts from British Shipbuilders, did all the design work, but it showed almost in cameo the frustration of British Shipbuilders. We had the right product, but we could not tender for it at the right price. We then, under John Parker pulled together a programme of ship designs, which was an effort to bridge the gap between standard production and the existing diversity of the yards. Later, although there were various official stances on products and on market research, it was really take what you could get, and the most attractive ships in many ways were the ones that you had not built before that were very complicated because then when you under-estimated the cost in order to get under the EEC ceiling, nobody really noticed. There was a totally unspoken, but implicit, understanding that when it came to the negotiating phase you pruned the estimate until you would sell the ship. Complex ships provided much more room for manoeuvre. We tended to look more competitive on complex ships, although in reality I suspect that was just because the estimating process was more nebulous.

In a sense, management consistently had the choice that you either close the whole thing as unviable, it was all losing money, this was a rational thing to do, but equally managers felt a loyalty to the industry, and there was a covert feeling that the job of British Shipbuilders was to protect the industry and to protect jobs. You had to come up with a compromise which would enable the government to save face by turning a blind eye, and letting the show keep on the road with the minimum disruption. Then you had to sell that to the unions.

My job as corporate planning director every year was to get forward a corporate plan which the government could accept, and would allow them to give us more money next year. That plan had to come up with some argument which dealt with the fact that you had failed miserably last year to deliver. It had to show improvement. It had to show some cost cutting, which meant closing down something or getting rid of someone. On the whole what would happen was that the yards would perennially send us a load of rubbish about their performance next year. They never did perform in that way because they were in a market where they could not perform.

We tried various techniques for devolving responsibility back to the yards. Robert Atkinson tried setting up divisional boards, but that did not work. So the next thing was to put in more managers. We tried to address the interface between Headquarters and the companies. Afterwards, two things happened. One is that Atkinson got overstretched. He was an old man by that time and was doing far too much with an impossible task to do. Then Graham Day came in and got rid of the divisional boards, and by that time we had introduced a quarterly monitoring arrangement whereby senior shipyard executives were accountable to the Headquarters Board. In the end, however, the yards were successful in keeping their autonomy. Some of them did all right, and others made an absolute cock-up. Mostly they just survived and got a little bit worse in a sort of suspended animation. I do not know what else you could have done. Like so many other things in British Shipbuilders, if you honestly analyse what is done, it is a casebook example, but it just did not work because the problems were so great.

The governments' support of shipbuilding may have been ineffective but it was certainly not inadequate. In fact I would say that basically the government leant over backwards to sort out British Shipbuilders and to assist them within certain parameters. It contributed £2 billion, plus the warship profits which for about four of five years went down the bottomless pit of merchant shipbuilding. I do not think that there can be any suggestion that the government did not make every effort to support the industry. If you are criticising the Conservative Government, it was probably for being too soft. They, on the whole, tended to pursue a policy which for most of the period consisted of the following: as long as British Shipbuilders did a bit of privatisation, and made some re-structuring to satisfy the EEC then they coughed up. It was not just subsidy upfront, but they continued to pay the unplanned losses afterwards. I do not suppose that any other country had more money really. We are talking about subsidy levels of £18,000 per man at one stage, and that, in my opinion is questionably good use of taxpayers' monies.

Interview No. 55: Richard Dykes, Director of Industrial Relations, British Shipbuilders Plc, 1977-1980

My first entry into the industry was in 1977. Up until then I had been a career civil servant at the Department of Employment, but was seconded to industrial relations at British Shipbuilders. I jumped at the chance to gain some industrial experience. Presently, my job was really the Director of Industrial Relations in charge of industrial relations policy with responsibility for negotiations, pay, conditions and later, redundancy policy. I was still technically on the books of the Department of Employment, but my salary was paid by British Shipbuilders.

From my end of it we were dealing in a day-to-day, hand-to-mouth way with the implications of the Incomes Policy of the Government. What we saw in the industry was this extraordinary set of 168 negotiating groups. Overlaid on top of this was Stage Three of the Incomes Policy, which set a ten percent limit on pay rises which could be got around through productivity schemes. I found myself dealing with the chief executives and industrial relations directors of the yards who were trying to explain to us the legitimacy of some pretty bogus productivity schemes to get round the incomes policy. We had to take all these deals to the Department of Trade and Industry. It was my first direct dealings of industrial relations, and I was astonished and horrified at the distance, running through the industry, between the management and unions. We counted some twenty-four unions in the Confederation that we had to deal with. Many of them were pretty politically motivated.

What we were trying to do was to devise a strategy, which gave us the best chance of settling the industry down, of getting more peace than strife, of trying to establish a framework of order on the industrial relations scene which would give us a better chance of improving productivity, and therefore of keeping our market share. We recognised that we were going to be driven to reducing manning, whether for efficiency reason or because the work just was not there, and so we had to devise a redundancy package that would enable us to slim the industry down without outright warfare. I think, in some ways, we were extraordinary successful in that this whole process could have gone badly wrong. We had a centralised wage negotiating structure within eighteen months. There is always something of appeal to the national officials of unions for the power to be centralised. It gave a chance for some of the unions to pull their members into some kind of order. Negotiations were spilt between blue and white collar, and I found myself negotiating on the white collar side [*draughtsmen, clerks, estimators, middle managers etc.*]. At one point, towards the end of negotiations, the issue turned on what rate we were going to pay tracers, who had a longstanding claim to be paid on a par with skilled draughtsmen. I did not know enough then to know that tracers were very far from being as skilled as draughtsmen. Eventually because it was the last point when the deal was either going to be made or not, I conceded that tracers should have parity of pay. Within a matter of a day of the deal being done, letters were received from the yards stating that what we had done on the tracers effectively contained the seeds of the deal's destruction. In actual fact the deal held, but the chief executives of the yards wanted it to fail because, in failing, it would demonstrate that the centre did not have a role. On the Union side John Chalmers of the Boilermakers was probably more responsible than anyone for making it happen, he and Ian Farningham who was the Managing Director of Industrial Relations really embodied the spirit of what we were doing.

Without a sensible redundancy package there would have been no chance of an orderly decline. We did research into one block of redundancy which affected 1100 workers in Swan Hunter, and we did research on what happened afterwards. We also did counselling to help them find other jobs and worked with the Manpower Services Commission. It was remarkable how quickly these redundant workers either left the unemployment register or never appeared on it in the first place.

On the question of whether nationalisation was correct? It depends on where you are standing. If you were someone with a vested interest on the management side, say Vickers of Barrow, where the order book was fairly secure, and where they felt that they were cheated by nationalisation, nationalisation was a bad thing. What they got was loss of power and independence. They could have carried on without nationalisation and might well have done better without it. They also took the view, which I do not agree with, that because their order books were secure they were by definition efficient. When you are operating on cost-plus contracts you have got to be deaf, dumb, blind and dead not to make money. They made money out of the taxpayer. They gave the impression that they were in the vanguard of the industry, but the facts did not bear that out at all. We found more sympathy for the merchant yards that were in competitive trouble. Some of the 'Spanish practices' were terrible. At Cammell Laird when I joined the industry there was something like 800 workers over the age of seventy, and there was a practice that you could clock in the morning, jump over the wall and go home, and the climb back in and clock out.

Mike Casey appointed me as a non executive director of Austin and Pickersgill. On turning up for my first Board meeting I got a pretty frosty reception from the Chairman, Derek Kimber. He saw that yard as his. He wondered what I was doing asking him personal questions about his business. From the government point of view they were fire-fighting to an extent. They owed some political debt to the unions, and the last thing they wanted in the run up to the 1979 General Election was warfare in the shipyards as well as everywhere else. As for the longer term I am not sure that their view was strategic for the country. The industry needed stability, and they could not afford to let it go to the wall. The only was they could see it surviving was in public ownership.

One thing that did strike me when I went around the yards was just how badly the workers were treated. I was in Scott Lithgow being driven around by a senior manager as the shift broke, and he wound down the window and shouted abuse at the workers. I remember vividly going to another yard and having lunch in the Board Room groaning with silver and white tablecloths, and then being taken around the yard by a director and noticing some people sitting somewhere eating sandwiches out of a box. I remember asking him if they provided canteen facilities. His reply was that I must be joking, as they could not afford to provide that sort of thing and in any event the men would not use it. Some of the yards were simply slums. We went across to Sweden at Kockums of Malmö and the place was a different world as the workers were treated properly.

Interview No. 56: R.J. Daniel, Board Member for Warshipbuilding, British Shipbuilders Plc, 1979-1984

My ambition was to become a member of the Royal Corps of Naval Constructors. I began an apprenticeship at Chatham Dockyard where I won a series of Admiralty prizes and at the end of four years I was given a Naval Constructors Cadetship and went of to the Royal Naval Engineering College. I then moved onto Greenwich, and then Bristol and left the Naval College there in 1942 and joined the Royal Corps of Naval Constructors and got involved in submarine designs, and later got the job of designing Britain's first nuclear submarine. In the end, I rose to head the Royal Corps as Director General Ships. I did this for six years until in 1979 I was asked by the Labour Government to join the Board of British Shipbuilders, and did so in April as Board Member for Warshipbuilding. After the General Election of that year we then had a very hostile period with Conservative Ministers whose sole aim was, in fact, the dismemberment of the industry. We prepared to sell off the warship sector. In 1984 I went down to Bath and set up Warship Design Services, which then became a part of Vickers. I then did a lot of marketing and advisory work for Vickers, retiring in 1991.

On nationalisation we found some extraordinary things. We found that when we tried to develop common salary scales across the industry that the warship yards were paying less, by and large, to their men than the merchant yards. We had one merchant yard which on closer inspection was found to be paying a productivity bonus of £5 per week to every employee from May to November 1979, and it did not have a single ship in the yard. The Managing Director of that yard left a week before we sacked him. What I missed when I joined the British Shipbuilders Board was intellectual dialogue. I found there was very little of that as people were overwhelmed by the day-to-day problems. By 1979 the unions had been brought substantially on board. This was largely due to the Confederation's Alex Ferry who was a tremendous chap. John Chalmers was also a reasonable man. The labour relations situation had improved markedly and at all levels there was consultation. There was a general desire to make the thing succeed. We were marching men out of the door in their thousands, but there were few strikes.

There was an enormous change after the election of the Thatcher Government in 1979. We were already rationalising the industry, running numbers down and closing yards, and the Conservatives demanded more closures. I have to say that I heard a Minister express pleasure when he was told that we had to close a yard. "The Prime Minister will be pleased," he said. They seemed to operate with little concern for the social consequences of these actions. To his eternal credit, Robert Atkinson fought like mad for the future of the industry. This attitude was not mirrored with the ministers. The other thing that we found was that the officials in the Department of Trade and Industry did not go out and look for loopholes in the Act which would help us wriggle through to keep people in employment and get orders. The Intervention Fund was implemented rigorously rather than sympathetically. We were looking for small subsidies for certain yards to keep orders coming and employment levels. We would be told that the Department would not be prepared to accept the implications of that loss, that £5 million there even though the total cost to the taxpayer of closure in redundancy etc would be much more. There was never a co-ordinated, cohesive policy on the part of government in the nation's interests. By contrast, the Dutch won an order from Taiwan to produce two submarines with the Dutch Employment Ministry paying for the cost of the hulls to preserve employment. We fought very hard to get warship orders into places of high unemployment to preserve their longer term shipbuilding capability, but the Department of Trade and Industry was not sympathetic.

In summation, I believe there was an inevitability about nationalisation as a result of having elected a Labour Government in 1945 and subsequently. The fact that shipbuilding was included in their remit of decaying industries that needed public ownership led to eventual nationalisation. You could argue it away in the 1950s and 1960s with full order books, but the writing was on the wall even then. I believe there was a certain grim inevitability about most of it. I believe that had there been a greater commitment by the Conservative Government to the nationalised British shipbuilding industry there would have been more pressure in Europe to establish some sort of tariff wall to maintain a bigger nucleus of indigenous shipbuilding. A government that was more outgoing in relation to Europe, and to the shipbuilding industry, might well have salvaged a bigger industry. That is my profound belief.

Interview No. 57: Robert Atkinson, Chairman, British Shipbuilders Plc, 1980-1983

I started my career during the Great Depression serving as an Executive Officer in the Merchant Marine, gained various certificates and joined the Royal Navy as a reserve officer, was promptly called up, and served right throughout the war with the Navy. During that time I had served in command of a tanker and six warships. After the war I was awarded a scholarship to the Massachusetts Institute of Technology in Boston, but did not take it up preferring to take an external degree at Imperial College, London. On graduating I went over to Canada and did a year's post graduate study in mathematics at McGill University before working in shipbuilding with the Davy Company of Quebec building all-welded ships with aluminium structures. I came back to England and became a director of Hamworthy, a marine auxiliary manufacturer of pumps, compressors and boilers. From there I became Managing Director of William Doxford, the marine engineers in Sunderland. There was a lethargic lot in charge who really behaved disgracefully. They lived the life of barons, the life of luxury, but did nothing intellectually or financially to push their country forward. I left Doxford, and went to Tube Investments, and from there to become chairman of Aurora, a steel and engineering group in Sheffield, and whilst there I was invited to become Chairman of British Shipbuilders. I then retired from that in 1983 and from Aurora at the end of 1984.

I had been six months in British Shipbuilders. The Permanent Secretary, Sir Peter Carey asked to see me to see how I was getting on. I laid down my divisional structure for the yards etc. He listened to me for about an hour, and then told me "Margaret [*Thatcher*] wants rid of shipbuilding." I still did not take the message in properly. So in my time no man ever said, what is your plan to improve strength and productivity? I had a new engine being developed under Peter Milne. We worked on it for two years and it was a winner. The day I left, Graham Day came in and with the approval of the Government stopped it. Norman Lamont [*Minister of State*] only seemed happy when I had closed down a yard or lay off 5000 men. At no time did Government ever sit down and discuss our size, let's have a business plan, let's rationalise, all they were interested in was closure.

In the early stages, nationalisation brought stability. I saw James Callaghan after 1979 and told him that he had done a great job with the Polish Order. He was surprised at that and told me that I was the only person who thought that. I think that the fact that he got orders from Poland, kept yards, which would otherwise gone under, in existence. He helped the industry. As soon as the Conservatives got in there was difficulties. They were always against shipbuilding, and they did not want the industry. When I resigned I said that I would not do a second term against the background of the threat to the industry [*privatization*]. I was not interested in keeping ship repair, but was adamant about keeping merchant shipbuilding and warshipbuilding as well as oil and gas rigs. The Government said they could not agree to that. They then brought in Graham Day.

In my time as Chairman I closed Caledon [*Dundee*] and that took a bit of doing because the men occupied the yard. I closed Henry Robb [*Leith near Edinburgh*], which I regretted because it was a good little yard. I closed Caledon in December and said switch off the electricity, we are not paying the telephone bills either. The men occupied the yard for three weeks and then came out. I told Alex Ferry of the Confed that I was absolutely opposed to the privatisation of the warship yards, and could not go back on that.

From the day Graham Day took over the Government destroyed shipbuilding. Govan is now in the hands of the Norwegians, but Norman Lamont was absolutely adamant about closure. The first Minister of State I worked with was Adam Butler, he would have rationalised, Norman Tebbit helped a bit, and he let me carry on with Smith's Dock. Lamont said that he would like a report and went to Rothschild and asked them to report on the future of the shipyards. Lamont would not give me a copy. He went one step further and asked Production Engineering Limited to report on engine building, which said that we would probably have to let go North Eastern Marine [*Clark NEM*] and keep engine building at Harland and Wolff in Belfast. I was shocked. All we needed in engine building was here in Britain, We had on the North East coast, Doxford, and Clark NEM, and Kincaid on the lower Clyde [*at Greenock*]. He would not accept that. My strategy was in total conflict with Government. Day's appointment was primarily political with yard closures at the centre of policy.

To say that British Shipbuilders was an unwieldy bureaucracy is absurd, because we did not have much at Headquarters. I closed one building. We had a sales department which kept tally, which was crucial on the marketing side. You can not run an organisation without some kind of central authority.

Instead of paddling my own canoe to make British Shipbuilders a success, I think that I could have been partly to blame for not forcing the government. I was going to call in McKinsey [*management consultants*], and the mistake I made was not to follow that through. It was very early on and I was talked out of it by another consultant. I should have commissioned a report on the future of the industry, and then gone to government with some recommendations.

Government had hostility to the existence of British Shipbuilders. They did not want to know about research and development. They did not want to know about engine development. All they were interested in was closure. I was wasting my time in the event as I see it. Summing it up, it was clear that the Conservative Government were determined to get rid of shipbuilding, and at no time was any plan allowed to interfere with their decision.

Interview No. 58: Graham Day, Managing Director, Cammell Laird, 1972-1975, Chairman, British Shipbuilders Plc, 1983-1986

My background was as a Barrister trained in Canada, specifically in the marine context. After eight years at the Bar, I went to work for Canadian Pacific, for whom in the international commercial arena I bought ships. At the time we were setting up an offshore bulk shipping operation, I ordered three relatively small container vessels from Cammell Laird at Birkenhead in 1968. On April Fools Day in 1970 we were told that the yard was about to collapse, because they were literally running out of money. I spent the next three or four months in Britain working with the Industrial Reorganisation Corporation to keep the shipyard alive so that I could get my three container ships, which we did. The following year I was asked to come and run Cammell Laird because the management really had not made much change. I later became Chief Executive Designate in the Organising Committee of British Shipbuilders, but resigned because of the lack of strategy, and then came back in as Chairman in succession to Robert Atkinson.

I came out of a quite carefully structured business in Canadian Pacific. In going into Cammell Laird, information, of almost any type, was virtually non existent. I went in August 1971, and they were on the eighth budget for that year. There was no grip on the business, no strategy. The next year we started to make money, and all of the time that I was there we made money. We had a product strategy which we were fortunately able to deliver. We instituted a corporate planning process. When I went into the organising committee for British Shipbuilders with much trepidation, after the Second Reading of the Bill before Parliament, I wrote around to each of the companies in the industry and asked for their current corporate plans. I received one; the plan which I had drafted for my successor at Cammell Laird.

It is wrong to say that there was a lack of investment, because money actually had been invested in the industry. The question is rather that the investment was not concentrated; it was sort of shared out. It was not investing behind excellence or emerging excellence. At Cammell Laird there was a group of middle managers, which by any reasonable standards were excellent. The top managers, most of who went, were not very good at all. The electrical capability in the yard was really outstanding [*a legacy of the Polaris nuclear submarine programme*]. We were able to do quite innovative things in our principal product area, which were petroleum product tankers. The plant was appalling. We were shifting steel in the stockyard with a little crane that was pre World War One. In the wet basin, for any heavy lifts, we had a single crane which was War Reparations World War One, which took about thirty-eight people to do a single lift. I have many other similar examples. It was Industrial Revolution technology to a large extent. Cammell Laird did suffer very much from lack of investment. By way of investment we had one building we called the Highbay, a two-storey office building, which we called the Polaris Hilton, but there was nothing fundamentally new on the steel side, in the pipe shop or wherever.

In industrial relations you were always trading pound notes across the table. You could not reach people in trying to explain what was required and why. I used to get angry and say, "I can get a job, can you?" Early in my time there I did a number of things arbitrarily. I would just ram things through and take the chances. I closed down the general engineering business, and physically moved a lot of things around the yard, and changed some of the working practices. Our total costs were high. It was not that we were paying high wages per se, but it was the productivity of the hour worked, and the equipment that supported it that gave us the high costs. In my time at Cammell Laird, we were making our deliveries and improving on them. However, we were still taking longer to produce all the ships that we should have.

Inadequate marketing was a polite way of saying that there was no marketing. Essentially there were salesmen. I recall Sir John Hunter being interviewed on television. He was asked what the state of the market was. He replied that the market was not very good, because they had not had many enquiries. There were some people who were nothing short of geniuses as salesmen, Ross Belch being one of them. I remember in around October of 1976 I went to Gerald Kaufman [*Minister of State*] and told him that the industry had fundamentally changed, not merely in terms of marketing sophistication etc, but because the *OPEC* [*Organisation of Petroleum Exporting Countries*] oil pricing initiative [*OPEC first tripled and then quadrupled the price of cruse oil in late 1973 and early 1974*] had changed the whole of shipbuilding. It is only now that some of those consequences are starting to flow through. He assured me that this was not the case and that it was only another cyclical downturn, because I was urging upon him the need for some very serious restructuring. It was then apparent that Sweden was going to get out of shipbuilding. They were going from number two to zero. Germany was going to sink. I used to say that there was the Swedish solution, which was to close everything down, or there was the German solution, which was to try and select a core and get to it as quick as possible. Politically [*in Britain*], the determination was to protect the perimeter, which meant all of it.

I believe in personal terms, that if I had been able to persuade the Labour Government and the trade unions in 1976 that a very radical restructuring of the industry in size, changes in working practices like the unions had never contemplated, focused investment, it would have had preserved an industry larger in Britain than that which exists today. I really believe that there was that single point in time which could have been decisive. In part of the unions there was a private acknowledgement of the realities. We had to try and convert that into a semi-public acknowledgement. There was also a need in the nationalised industry to radically change the management. I wanted to do a Cammell Laird on British Shipbuilders.

I am not drawing a political distinction here between the Labour and Conservative Parties, because during the passage of the Bill the Tories behaved appallingly, but the whole thing was so politicised that there was no industrial hope in my view. If there was a failure in nationalisation it was that the government of the day did not have a strategy for the industry. They had a policy which was nationalisation, and I for one thought that the industry was in extremis. When I was asked to try and manage the industry, I believed that nationalisation theoretically provided a framework within which something could have been done with the industry, and I very rapidly realised that there was no strategy for the industry at all. Worse than that, was that nobody was interested in the government of the day in having a strategy developed.

I believe that nationalisation was totally political at the outset as indeed I believe that privatisation was totally political at the outset. There were two different ideologies. In my innocence, I believed that there was something behind the nationalisation ideology. I was told there was a deal done with Danny McGarvey of the Boilermakers, or that there was a deal done with somebody else. Nationalisation gave rise to some positive things, but for the bulk of the industry it almost froze it in time.

When I came back in 1983 things were almost in a time warp. One of the things which had happened was that the financial information systems on a company to company basis which were appalling in 1976 were actually very good in 1983. There was considerable strength in the finance function. There was a development of technology and systems, which if it had taken place in the 1950s and 1960s the industry would have been in a very much better state. It was sadly just too late. In working practices and the general quality of the management not much had changed. The then government's policy was to privatise, but for very pragmatic reasons. Firstly, the industry was costing the Government a lot of money, and was an inefficient way to spend government pound notes. Secondly, also, unless you could break out some of the businesses which were likely to survive, then you were likely to lose all of them. Some of these were the warship firms who are now facing some kind of new reality. Prior to nationalisation they were probably the least competitive. They were the least efficient, the most overmanned, and the worst managed.

I think the industry would have survived better without nationalisation if the government had let market forces deploy. Equally I agree that if government had stood back and said that intervention was not for them there would have been an accelerated move towards market realities.

If you ask if British Shipbuilders understood on a global basis what was happening to the market place, then the answer is yes. Market research, due to Dr. Martin Stopford, had improved. He provided hard data with intelligent analysis and pointed people to face the issues. Personally, I think the marketing was extremely uneven. There were great conflicts as to whether the centre marketed and the yards sold, or whether the yards marketed and sold, or whether they both marketed and sold. The function of the centre was grossly overmanned. I bought three ships from Govan from Canada, and it took me nine months, finally going to Eric Mackie privately to try and order the ships. I could not get anyone in British Shipbuilders to take me seriously.

British Shipbuilders had developed into a bureaucracy, because it was led by people who had either no commercial experience at all, or if they had, saw that they had a position. There was a tremendous amount of self-aggrandisement. It was a vast and very expensive bureaucracy.

After 1983 I agreed, through a two page letter and a timetable, privatisation plans with the government, and given that both sides kept their ends I have no complaints. We were on the verge of a national strike in 1984, but it did not happen for two reasons. Firstly, the union was finally convinced that we had said what we meant. Secondly, they understood the economics. It was realism.

It was an industry which had such a marvellous history, and at times of national crisis had become very important. It never wanted to see itself as things really were. They could not even see the untidiness in shipyards. They were not even wearing rose-tinted glasses; they were opaque. There was unbelievable optimism that the market would pick up. I remember one great phrase that, "we should build in advance of demand." They did not want government interference but they wanted their money. We moved from hearing their pleas from an industrial basis to a social basis. So the social case transcended the industrial case. There were no penalties, and there were no sanctions.

Britain is finished as a maritime power. I do not know if this was inevitable.

Interview No. 59: Geoff Fuller, Board Member for Warshipbuilding, British Shipbuilders Plc, 1983-1986

I joined the Royal Navy in 1944, training as a marine engineer and then moving over to naval architecture. I completed my training in 1952, and from then until 1983 I worked for the Ministry of Defence, mostly on design, concentrating on submarines, but including small craft, frigates and the management of ships materials, and computer projects. Towards the end of my period I was the Refit Project Manager for nuclear submarines, including the research and development side of submarine warfare, and my final two jobs were Director of Naval Ship Production, the interface with the industry, and then for about a year, Director of Productivity and Industrial Relations for the Royal Dockyards. I then joined British Shipbuilders as Board Member for Warshipbuilding in 1983 and was with them for three years, and that included ancillary jobs such being Chairman of the Council of the British Ship Research Association, and also Director of Technology, and head of their training company. Towards the middle, due to the problems at Vickers, Barrow, I was sent by the Chairman of British Shipbuilders to take over the running of the Barrow shipyard for the final two years before its privatisation.

The industry was broke before it was nationalised. It eventually capitulated, and thereafter, the individual owners wanted to stay on the gravy train, and if they could keep their heads down and keep their firms un-violated by these horrible people at headquarters then they might survive to be re-privatised. So from their point of view nationalisation was a wise policy. From the national point of view I am not so sure that once it was nationalised it would not have been better to have kept it together and ended up with a privatised British Shipbuilders with two or three yards who had a chance of competing reasonably well, and I include that in the merchant side. I do not think that the industry would have rationalised itself. The 1970s created the scene for nationalisation to do it. In the 1980s if it had not been done then market forces would have dealt with it and you would have ended up with few yards, if any.

British Shipbuilders management in the beginning was fairly inept. I do not think the first Board was all that competent. They had problems; they did not deal with the industry quickly. They allowed themselves a lavish top organisation with confusion between federalism and corporate ownership. I suppose if they had put in a very powerful boss at the time, Ian McGregor [*Mrs. Thatcher's choice to rationalize the British coal industry*] or someone like that, he would have fired the lot and started again, and it would probably have been a pretty good industry.

The warship firms saw the government in the shape of the Department of Trade and Industry and the Ministry of Defence as two separate animals. I think they saw those who nationalised as quite separate from their "real" customer, the Ministry of Defence, and in particular, their customer, the Royal Navy. Most of them had built up extremely close relationships with serving naval officers. There was a very close relationship between the Navy, the shipbuilder, and to some extent, the naval side of the Ministry of Defence who were really totally separate from the objectives of the Department of Trade. The fact that the Ministry of Defence as contractee would insist that tenders from member companies of British Shipbuilders were not communicated to British Shipbuilders corporate headquarters was indicative of that type of outlook. The Ministry of Defence did not see nationalisation as meaning anything to them. Looking back on it that is the view that I have come to over the last few years. They dealt with the yards quite separately as they had done before. Personally, I can recall being accused by a Director of Contracts [*at the Ministry of Defence*] of knowing the results of tenders when I was at British Shipbuilders, and that this could be a criminal charge. What they failed to see was that if they got an unduly cheap ship from a shipbuilder, he would go into a loss, which British Shipbuilders corporately had to fund from the Government.

I suppose that the Ministry of Defence would look on it as a way of getting more money out of the Treasury, although I am sure that they would not say that overtly.

On privatisation, I think it was destabilising for a time because each firm did not know what it was doing next, and the managements' were involved in management buyouts. The big error in privatisation was that it should have been done as one package and all firms should have been sold on the same day. What happened was that because some firms went at different times, different groups ended up with firms they might not necessarily wished to have had. It would have been a far better situation if there had been a single date so everybody was on a level playing field. Whoever was leading a management buyout was still being paid by British Shipbuilders at the same time. It is a pity that there was not more outside bids. It would have injected a new management concept into the industry.

The money that went down the drain in British Shipbuilders was due to extremely poor management in the first five years. At the top, initially, were Casey and Griffin who were treating it as an industry which was not sick even though order books were collapsing. Atkinson just held the ring, and really refused to cut numbers and it was not until Graham Day arrived in 1983 that management got a grip on things. As a result, £1.8 billion of taxpayers' monies went down the drain, making the present government totally and completely bored with shipbuilding. They look on the shipbuilders as a group of whingers and "cap in handers," and they are not about to save any shipbuilder at all. The other thing that they do not understand is the desire on newly industrialised countries to use shipbuilding as a trigger for industrial development, even if it cost money. We do not need that, Korea did it.

Shipbuilding has been a fascinating industry to be in. Shipbuilding is a different world. The building of a ship has to be done with care, although in many ways it involves low technology. It is a huge piece of kit, and it is much more akin to building a power station or similar large artefact. It is an unusual industry and a difficult industry to understand. But, as a final point, do not forget that a large part of the western world shipbuilding industry no longer exists. The Swedes do not build any merchant ships, and they were the leaders forty years ago. Germany still builds some. Japan still produces fifty-three percent. We are not alone. We are just more emotional about it.

Interview No. 60: Maurice Phelps, Director of Personnel, British Shipbuilders Plc, 1980-1987, and Chief Executive, 1986-1987

In 1980 I was with Leyland Vehicles when Robert Atkinson phoned me up and invited me join British Shipbuilders as Head of Personnel. I joined in December 1980 and went through three years with Atkinson. Then Graham Day joined and I worked very closely with him from 1983 to 1986. Graham left in

1986 to take over Leyland Motors, and Philip Hares took over, but had a heart attack at the end of 1986, and I was asked to take over as Chief Executive, which I did. I left in May 1987 and joined Sealink [a ferry operator] staying with them to August 1989, when I left to start up a joint consultancy firm.

The weaknesses for a newcomer to the shipbuilding industry were difficult to see, but they were there. Firstly, there was a high level of traditionalism. Secondly, management was highly introspective; it was very difficult to bring in anyone from outside the industry. The third problem was undoubtedly the trade unions. The industrial relations structure that was set up produced peace and stability, but it did not produce productivity and change. So what was done, to produce one negotiating body which had control from the centre, had the effect of actually producing a reduction in the number of major disputes. Therefore, we had more man hours in which to produce ships, but in my view the total effect of all this was to slow down the process of change. When I arrived at the end of 1980 there was a whole atmosphere of consensus, of joint ownership between the Trade Unions and the management, which reached the stage where nothing could happen in virtually anything unless there was joint agreement.

What I would have preferred to have happened is that they would have forced more changes in flexibility, interchangeability, productivity and through this produced more conflict. I think that the unions were in such control of the situation that the middle managers had lost their guts. I think that the trade unions were enormously powerful. They had marvellous leaders in people like Chalmers and Alex Ferry, who I still believe was the best union negotiator I ever encountered, and who had a perspective not many had. So they had men of true size and stature, but they had also as a result of the way that the industry had been shaped, and I think the very individuals who they had put in charge of the industry in its formative years of nationalisation, a situation which prevented real changes taking place.

When I arrived there were two agreements in place, which when read amounted to very little. I was condemned for that. There were productivity clauses with good intent, but when you actually looked at the evidence then nothing had actually been done, and no changes had really been made to improve productivity. The big claim was that somehow they had miraculously stopped the major disputes occurring and that they had reduced their numbers. That was not what it was about. Two weeks after I joined we declared 1200 redundancies, which had to be voluntary because compulsory redundancy was not tolerated by the trade unions. We spent a considerable amount of time with the trade unions agreeing that we were actually going to reduce the number of people. I experienced an almost unhealthy closeness, and I suspect that there was not much that we discussed on the Board was not already known by the unions.

We struggled to put through real changes in working practices in the first two years. Then in 1981-1982 the Chairman, Robert Atkinson embraced the Productivity Improvement Programme, which he largely translated into a computerisation process. As a result we spent something like £40m on new technology, which had very little effect, in my view, since it never really paid off. What it enabled us to do was to lift our eyes and look at the competition. In order to reach Japanese levels of productivity we would have to increase ours by something like fifty percent. This lag was purely down to management of working practices, organisation and procedures. It disguised a high level of excess capacity in the UK industry. Eventually we began to come to grips with it at the end of 1983, and then through 1984 and 1985, recognising that we would have to reduce capacity and only maintain those yards which were of potentially high efficiency. Eventually we did that in 1986-1987, but I think it was wrongly done because we fell into the trap of believing that if we consolidated historically individualised yards, then somehow we would save overheads.

At the end of his period in office, Atkinson took a battering from the Government because they did not think he was moving fast enough [on privatisation]. One saw at the end of 1983 a much tougher line being taken by the government on where the industry was going. The clearest example of this was the desire to privatise the industry. Graham [*Day*] was in a sense, a breath of fresh air. He did not conceal, in any way, where he stood. He said quite clearly to the trade unions that he would never intentionally deceive them. He told them that the warship firms were to be privatised within a certain period of time. That was the job of the business. Graham set out to achieve a particular objective which had been set by government and he delivered it to the day and to the letter. The difficulty, of course, was running the business when you were pulling fifty percent out. The danger to the business, which was quickly revealed, was that if you pull the profitable part out, you leave behind the unprofitable part in all its unholy colours. Although the warship firms were profitable, there is no evidence that they were any more efficient than the commercial shipbuilders. They might even have been less efficient. There was less incentive for them to become efficient because they were profitable. If you improved warship productivity the Ministry of Defence would cut the price of a ship, so there was little incentive to do so.

There was a very interesting period at the beginning of 1987 at the time Philip Hares had a heart attack, we were at that time trying to negotiate orders for Govan in China, and we had just launched a new marketing strategy on the basis of a reduced capacity for merchant shipbuilding. Government at that time began to throw money at us to get the orders at all costs because the General Election was coming up in May 1987, and they were anxious that they should not weaken their Scottish position any more that it was going to be. All the indications coming through were that two things would happen, post election, if the Conservatives got in. Firstly, they would privatise the rest of the industry very rapidly, or, secondly, they would close it down. They got back in and John Lister was appointed Chairman and Chief Executive. He maintained that he did not understand that that would be the Government's position. He believed that he could actually do a rescue job, and produced a corporate plan which said that by centralising sales and marketing he could keep running NESL and Govan.

Kenneth Clarke [*Minister of State*] met the Board and made it very clear that he did not see shipbuilding continuing in the nationalised form. Rather that to kid the workforces anymore that there was a future in shipbuilding he would much rather come out, up front, and tell that, that for example, in the North East, there was no future in shipbuilding, and that they were going to get money from the EEC to revitalise the area, and bring new industry into the yards. It was very clear that after those meetings the future of British Shipbuilders was pretty well gone. John Lister found himself in a position of dealing with a government who were very determined to privatise and close down shipbuilding. I suspect that they did do a deal with Europe, because a certain amount of aid would be given to produce new industry if there were be no further shipbuilding in the North East.

Conclusion
The Decline of British Shipbuilding in Context

The rationale of this book has been to let the actors speak for themselves from a position of informed and measured opinion. From the contributions offered, the reader can readily discern patterns in the decline of the British shipbuilding industry and to a lesser extent that of marine engine-building and ship repair. In the conclusion we do not purport to analyze any contribution in detail except to give the reader some context into which to place the decline.

Prior to 1914, the British shipbuilding industry dominated world output, regularly accounting for around sixty percent of the world's ships. At the same time, the British mercantile marine was easily the world's largest trading fleet. The Royal Navy stood as the world's foremost navy, and Britain's combined naval and mercantile strength, inextricably linked to the Empire, made the nation the world's foremost maritime power.

The Great War and its aftermath radically altered Britain's position. Losses to U-boats, mines and other countermeasures were severe. After 1918, port congestion, the need to replace tonnage lost and a short-lived speculative boom that ended in 1920 when freight rates collapsed placed considerable strains on British shipbuilding, whose capacity had increased by forty percent during the war.[1]

Interwar shipbuilding was characterized by excess capacity relative to demand. There were two real slumps in the early 1920s and early 1930s, a concomitant collapse in naval demand owing to international treaty limitations in place from 1921 to 1936, unemployment rates averaging thirty-two percent of the insured workforce during the 1920s and the rise of international competition fuelled by economic nationalism which led to uneconomic prices. British shipbuilding began its own rationalization scheme under National Shipbuilders Security Limited in 1930, and during the decade about one-third of the industry's capacity - first, entire shipyards, and then later, berths - was sterilized against any return to shipbuilding. Two great mixed naval and mercantile yards, Beardmore at Dalmuir on the Upper Clyde, Palmer's at Hebburn and Jarrow on the Tyne closed for good, with the latter leaving an unemployment rate of seventy percent in its wake. Profits were to a large extent restored, however, to the remaining mixed naval and mercantile firms - the largest ones in the industry - due to rearmament.

[1]In the industry's view capacity was the maximum output achievable with all berths utilized in any year given sufficient demand and with no supply constraints.

Mercantile demand, in stark contrast, remained depressed up to the advent of the Second World War. Merchant shipbuilding generally lacked investment in new technology: the tramp sector suffered most, and owners largely remained wedded to steam rather than motor propulsion. Management was old, as was the workforce, and those that remained after the interwar chaos saw shipbuilding as their main occupation. There is no doubt that at the outset of World War II, British shipbuilding was both quantitatively and qualitatively weaker than it had been in 1914.

The Second World War quickly eradicated the problem of weak demand and uneconomic prices, and it laid the basis for an unparalleled period of prosperity for the industry which lasted for almost two decades. British shipbuilding was, however, not as well placed to take advantage of the long boom in international trade despite that the fact that it seemingly had many advantages in the immediate postwar period. The United States, for example, had sold off a sizeable proportion of its huge fleet of War Emergency-built ships to foreign owners, kept the rest as part of a strategic reserve and in so doing prevented the market from again becoming glutted. Thereafter, the US shipbuilding industry again retreated behind a barrier of protectionism and high costs. The economies of Germany and Japan had been severely dislocated, and both were initially subject to severe Allied restrictions on shipbuilding. Britain, in stark contrast, stood at the threshold of a seller's market, with shipowners eager for tonnage at almost any price.

In this booming market too little rather than too much capacity relative to demand now seemed to be the norm. By 1948, most of the industry's re-conversion work had been completed; Britain's output comprised fifty percent of the world total, and shipbuilders could now compile long order books. Shortages of materials, particularly steel, had an impact, but the extent of the problem has yet to be satisfactorily quantified. We do know, however, that the flow of orders generally was far in excess of what the industry could produce. Noting the size of Britain's order book, shipowners began to look elsewhere for quicker delivery. Although annual output in Britain held steady during the 1950s at around 1.3 to 1.4 million gross tons, this was in stark contrast to what was happening elsewhere: world output over the decade trebled to over nine million gross tons, and the world order book grew to twenty-seven million gross tons.

British shipbuilders had historically exported about twenty percent of its output, instead relying principally on domestic owners to fill its order books. Capturing a greater proportion of an expanding world market for non-British ships during the 1950s was hardly part of the industry's equation; instead, the retention of bespoke linkages with domestic shipping firms remained paramount. The year 1956 was an *annus horribillis* for Britain on two fronts. First, it was the year of the Suez debacle where Britain finally lost any pretensions it still harboured to great power status; and second, it was the year in which the British shipbuilding industry lost its number one place in output – and hence the title of the world's premier shipbuilder – to Japan. Two years later, Britain fell to third place behind Germany, and the country's share of world output had plummeted to fifteen percent.

The years after 1958 were increasingly difficult for the industry as freight rates dipped and the long-term sellers' market came to a jarring close. Order books shortened, competition intensified and the major growth markets – large tankers and bulk carriers – largely passed Britain by. Industrial relations, if one could dignify what went on in shipbuilding with that term, remained volatile. The industry remained largely casualized, dominated by hire-and-fire policies. Demarcation between trades still held sway, and inter-union rivalry was fierce.

During the 1950s and the early 1960s, the industry's trade association, the Shipbuilding Conference, regularly approached the Conservative government about increasing the volume of naval tonnage, advancing naval programmes and facilitating credit guarantees so that domestic shipowners could place orders in British yards. At first the consensus on how to reduce the looming tonnage glut was to argue that countries other than Britain should reduce capacity. But faced with weak order books, poor prices and slim profits, the Conference soon approached the government with a plan to reduce domestic capacity voluntarily. Although this was a protracted process, it was ultimately unsuccessful, and no scheme on the lines of the 1930s ensued. By 1963, however, the Conservative government did put in place a one-year Shipbuilding Credit Scheme in the amount of £75 million. Although this scheme had some success, it masked the underlying lack of competitiveness of the industry in relation to Japan, Sweden and Germany. Costs in Britain were too high and productivity was too low, factors that would bedevil the industry for years to come. Although some form of concentration in the industry was discussed, in the end it was rejected and did not re-emerge until after the election of a Labour government with a working majority of five in October 1964. Under Labour, the Board of Trade sponsored an independent Shipbuilding Inquiry Committee chaired by Reay Geddes.

The ensuing decade was one of the most remarkable in the history of shipbuilding. World shipbuilding output soared from ten million to nearly thirty-six million gross tons, and by 1975, after the first OPEC oil shock, world capacity was in the region of forty-five to 50 million gross tons. Japan held half of the market, and one ship type, the tanker, regularly accounted for half of all orders. British shipbuilding, however, sagged, struggling to produce an annual output of just over one million gross tons per year. The nation slipped relentlessly down the league table of shipbuilding nations and sank into a crisis, albeit one which eventually engulfed most European shipbuilding nations.

The Geddes Report, which was accepted almost immediately by the Labour government in March 1966, offered panaceas which were very much in keeping with the government's wish list for the industry: improved efficiency, lower costs and increased international competitiveness. The Minister of State for Shipping, Roy Mason, had visited Japan and Sweden (as had most members of the Geddes Committee before producing their report) and came away believing that there was a need for some form of concentration as well as a separation and rationalization of marine engine-building capacity. The industry's previous arguments on capacity received short shrift, as did the unions' wishes for nationalization. Instead, Geddes recommended grouping firms on river centres to reap economies of scope and scale, such as specialization, improved bulk buying power, better productivity and improved marketing and industrial relations.

Geddes recommended five major groups; not more than two on the Clyde; one encompassing Harland and Wolff, Vickers [Barrow] and Cammell Laird; one on the Tyne; and another on the Wear. Overseeing the administration of the process would be a newly created Shipbuilding Industry Board, which was *in situ* by July 1967 and was empowered by statute to give grants and loans to aid the process. In the event, two groups emerged on the Clyde, geographically split into Scott Lithgow on the Lower Clyde and Upper Clyde Shipbuilders; one on the Tyne under the leadership of Swan Hunter and including Vickers' High Walker yard, Redheads, Hawthorn Leslie and Wallsend Slipway; and one, eventually, on the Wear with Austin and Pickersgill and Bartrams. Vickers (Barrow) and Cammell Laird (both heavily involved in nuclear submarine construction) remained separate. Harland and Wolff refused to merge and remained independent.

A devaluation of sterling in 1967 improved Britain's price competitiveness, and the closure of the Suez Canal due to the Arab-Israeli War stimulated orders for big tankers. Ever-larger crude oil tankers made the long haul from the Persian Gulf to Western Europe via the Cape of Good Hope economic. Accordingly, the world order book more than doubled between 1967 and 1971. Even the British shipbuilding industry participated in this boom, as the size of the order book doubled. Employment was maintained, and the industry benefited from government aid of just under £160 million. Nevertheless, beneath the surface there was a serious deterioration in the industry's financial health, particularly on the mercantile side which incurred heavy losses, reductions in net assets and falling amounts of working capital. Liquidity, the most important factor in shipbuilding, was squeezed; late deliveries of ships were common; and contracts taken at fixed prices in 1968 and 1969 suffered the effects of a doubling of the rate of inflation from three to six percent.

British shipbuilders then returned to their familiar litany of complaints about unfair foreign competition, tariff barriers and subsidies, particularly in the case of Japan; closer to home, the industry railed against restrictive trade practices which restrained moves toward greater flexibility and interchangeability in the workforce. Growing gaps in order books resulted in overheads not being fully funded and drained reserves to cover falling profits. By 1970, both Cammell Laird and Harland and Wolff, two of the big six British shipbuilding companies, were in effect wards of State. In 1971, Upper Clyde Shipbuilders all too predictably imploded which allowed Yarrow to escape back into the protective arms of the Ministry of Defence and hence to continue in business. Three of its former partners – John Brown, Fairfield and Stephen – had prior to 1968 been mixed naval and mercantile firms, but now Yarrow was the only warship firm on the Upper Clyde. John Brown was eventually sold to Marathon Oil, Stephen closed and the old Fairfield yard became Govan Shipbuilders and was effectively nationalized. On the Lower Clyde, Scott Lithgow struggled on, but the effects of fixed-price contracts on super tankers had already sown the seeds of its destruction. On the Tyne, the Swan Hunter group was in similar difficulties.

By 1973, the publication of a report by the consultants Booz-Allen and Hamilton entitled *British Shipbuilding 1972* put the industry's competitive position into sharp focus. All the usual suspects were identified – poor marketing, a multiplicity of unions, uncompetitive prices, late deliveries, poor productivity, management failure, lack of capital investment and a failure to avail of new technology. It was strikingly similar to the Geddes Report seven years earlier except that Booz Allen did not attempt to inspire confidence, opting instead to tell it as it was. Britain, alone among major shipbuilding nations, had experienced no growth in recent years; output had remained static, and shipowners, domestic and foreign, were reluctant to place orders with British yards. At most levels of government support hypothesized in the study, employment would decline by 30,000 to 40,000. Booz-Allen recommended that warship building be concentrated in three specialist yards, Vickers [Barrow], Vosper Thornycroft and Yarrow, eventually leaving the three remaining mixed and mercantile builders, Cammell Laird, Scott Lithgow (effectively conventional submarines) and Swan Hunter out in the cold.

Predictably, the shipbuilders, with the exception of the three specialist warship firms, were not enamoured of Booz-Allen's recommendations. Already in 1972, the Conservative government under the Industry Act had introduced a three-year programme of declining construction subsidies. Although the Booz-Allen report should at the very least have led to a radical overhaul in both industry and government thinking, its thrust was somewhat mitigated by a rise in oil freight rates which led to another ordering boom; by the end of 1973, British shipbuilding had its largest-ever order book amounting to 7.5 million tons. Unfortunately, this bubble of optimism was soon burst in the wake of the OPEC oil price hike: only 900,000 tons were added to the British order book in 1974 and a pitiful 73,000 tons in 1975.

A year earlier, another shipbuilding, repair and engineering group had been effectively nationalized in the wake of the collapse of the Court Line. As a result, Doxford and Sunderland Shipbuilders, Doxford Engineering and various ship repair firms came under government control. The drastic reversal in the industry's fortunes coincided with the return to power in 1974 of a Labour government already committed to nationalization. The industry and its friends in the Conservative Party furiously opposed this idea in Parliament and fought tenaciously against the Bill which barely passed in April 1977. On the one hand, the industry vigorously opposed nationalization, but on the other it sought short-term government aid. The industry had no strategy except opposition to nationalization and the vague notion that capacity should be kept in line with domestic demand through market forces rather than government intervention. On the eve of nationalization, a large part of the industry was rapidly running out of work and in danger of imminent collapse. The government's commitment was clearly to preserve employment; after all, it had already made this clear by effectively nationalizing a substantial proportion of the industry. Inefficient yards were kept in operation by government largesse, but this aggravated the competitive situation both domestically and internationally. The grouping of firms had clearly not been a success and had not produced world-class firms with clearly coordinated strategies and realizable long-term objectives. Plainly, the majority of firms were hanging on by the skin of their teeth, and by Vesting Day, 1 July 1977, the majority of the industry was barely alive as it limped over the starting line of nationalisation.

That the Bill had taken so long to be enacted was due almost entirely to ideological opposition to nationalization by the Conservative Party, which was also against the nationalization of the aircraft industry, which had consumed a larger proportion of public monies than had shipbuilding. That the two separate industries were combined in one Bill exacerbated the delays; given the rising tide of inflation, delays ate into firms' liquidity and profitability, precluding development and investment.

Ever since the 1930s, the private industry had consistently advocated a policy of capacity reduction, first through volunteerism in the 1930s and latterly by market forces. By July 1977, the industry, including the warship firms which by this time were effectively wards of state in any event, had reached the stage where its continued survival depended entirely on the ending of private ownership and its replacement by public control through a state corporation, British Shipbuilders Plc.

What was evident, although barely voiced at the time, was that more-or-less the same people who had put the industry into its present mess were left in control in the federalist structure of individual cost centres initially adopted by British Shipbuilders. It was clear, however, that the industry had to be brought under proper financial control from the centre, and this imperative soon drove policy, as did an orderly reduction in capacity enhanced by voluntary redundancy schemes from August 1978. British Shipbuilders faced a market situation hitherto unparalleled in its severity, and one it did well to survive against seemingly horrendous odds. In its first nine months, British Shipbuilders posted a pre-tax loss of £108 million even after receiving Intervention Fund assistance. The Polish ships deal, financed by borrowing US $65 million by issuing bearer bonds with a coupon of nine percent repayable between 1987 and 1992, although hopelessly uneconomic at least gave the Corporation some breathing space, as did the continued application of the Intervention Fund with European Economic Community approval.

By January 1979 a national agreement on wage and salary structures had been negotiated with the trade unions, and by the end of the reporting year trading losses had been stemmed at just under £50 million. Nonetheless, British Shipbuilders was chasing a moving target as the market continued to contract. The total world volume of new orders was a mere eleven percent of the 1973 peak, constituting perhaps the worst commercial climate yet known. Competition was accordingly severe, prices uneconomic and subsidization rife. The British industry was clearly engaged in a fight for survival, and restructuring was becoming increasingly vital. By the third report of the Corporation in 1979-1980, the Chairman, Admiral Griffin, and the Chief Executive, Michael Casey, had retired. The new Chairman and Chief Executive, Robert Atkinson, took over in July 1980 with a loss of £109.9 million posted and further capacity reduction an imperative. By the end of the 1980-1981 reporting year, Atkinson had introduced a divisional structure with five trading divisions, each with its own managing director and board of directors with responsibility for the constituent companies. These were merchant shipbuilding, warship building, engineering, ship repair and offshore. The latter, comprising Scott Lithgow and Cammell Laird, proved to be an unmitigated financial disaster. Yet by March 1982, the Corporation had posted, despite a contraction in the Intervention Fund, by far the best results in its short history with a trading loss of £19.8 million. The results were aided somewhat by the on-going weakness of sterling and the continued closures of yards.

By the time Atkinson retired in 1983, to be replaced by Graham Day (rather ironically given his resignation from the Organising Committee of British Shipbuilders) as Chairman and Chief Executive, the trading loss had reached a record £161 million, of which almost £100 million was attributable to losses in the Offshore Division alone. Privatization of the warship firms was now firmly on the agenda and was pursued unequivocally. Scott Lithgow ironically was both the Corporation's biggest loss maker and its first constituent firm to be privatized when it was sold to Trafalgar House Plc in March 1984 for £12 million. Work towards the privatization of the warship firms proceeded, and all other activities were concentrated into a single division. By March 1986, the warship firms had been privatized. Day resigned in April and was succeeded by Philip Hares, who retired in April 1987 due to poor health and was in turn succeeded by John Lister. To Lister fell the unenviable task of selling off the rump of the Corporation, which now consisted of Appledore Ferguson in North Devon and Port Glasgow, Govan Shipbuilders, the engine builders, Clark-Kincaid at Greenock and North Eastern Shipbuilders Limited (NESL) comprising four yards in Sunderland. By June 1988, Govan Shipbuilders had been privatized and sold to the Norwegian conglomerate, Kvaerner; by the end of the year, the last remaining large merchant shipyard, NESL, had been earmarked for closure and Appledore Ferguson and Clark Kincaid had been sold. British Shipbuilders was left with only NESL, which was soon closed. By this time the British shipbuilding industry in output terms had long been statistically irrelevant. The denouement of volume shipbuilding in Britain had been accomplished *de facto*, and there is little doubt that the industry's problems were so great that this would have occurred much earlier had it not been for nationalization and the borrowing powers afforded to British Shipbuilders by government, which by December 1979 amounted to a limit of £500 million; by March 1987, the borrowing limit had been raised to £1800 million.

Readers, academic and lay alike, can of course come to their own informed conclusions about the demise of the industry. What this collection of interviews does is to make a specific contribution to the existing historiography by highlighting the insiders' views, most of which are given with the assurance of hindsight. The views propounded in the text are, just as in life generally, undoubtedly affected by biases of one form or another and shaped by a certain amount of *post hoc* justification. What is evident, though, is that there is a clear cleavage of opinion between the warship firms and the rest, particularly on the formation and continued operation of British Shipbuilders Plc. The inclusion of the interviews of civil servants and those responsible for British Shipbuilders offers the reader a clear contrast to assess the weight of differing opinions. Finally, it is in terms of assessment that this book is particularly important, not just of itself but as a corollary to the existing literature, and it is in this spirit that it has been undertaken.

Select Bibliography

Books: Clyde and Scotland

Alexander, Kenneth J.W. and Jenkins, C.L. *Fairfields: A Study of Industrial Change*. London, 1970.

Ayre, Sir Wilfrid. *A Shipbuilder's Yesterdays*. Aberdour, 1968.

Bellamy, Martin J. *The Shipbuilders: An Anthology of Scottish Shipyard Life*. Edinburgh, 2001.

Borthwick, Alastair. *Yarrow and Company Limited, 1865-1977*. Glasgow, 1977.

Broadway, Frank. *Upper Clyde Shipbuilders: A Study of Government Intervention in Industry - The Way the Money Goes*. London, 1976.

Bruce, Stanley A. *Hall Russell Remembered: Shipbuilding in Aberdeen, 1864-1992*. Banff, 2009.

Buchan, Alasdair. *The Right to Work: The Story of the Upper Clyde Confrontation*. London, 1972.

Campbell, Roy H. *The Rise and Fall of Scottish Industry, 1707-1939*. Edinburgh, 1980.

Carvel, John L. *Stephen of Linthouse: A Record of Two Hundred Years of Shipbuilding, 1750-1950*. Glasgow, 1951.

Dick, Michael. *The 4.15 to Cartsdyke! A Tale of Two Shipyards - Scotts and the Greenock Dockyard*. Edinburgh, 1993.

Foster, John and Woolfson, Charles. *The Politics of the UCS Work-In: Class Alliances and the Right to Work*. London, 1986.

Herron, Frank. *Labour Market in Crisis: Redundancy at Upper Clyde Shipbuilders*. London, 1975.

Grangemouth Dockyard. *Grangemouth Dockyard Company Limited, Grangemouth: Tides at Dockyard*. Grangemouth, 1953.

Grant, Sir Allan. *Steel and Ships: The History of John Brown's*. London, 1950.

Hume, John R. and Moss, Michael S. *Beardmore: The History of a Scottish Industrial Giant*. London, 1979.

Johnman, Lewis and Murphy, Hugh. *Scott Lithgow: Déjà Vu All Over Again! The Rise and Fall of a Shipbuilding Company*. St. John's, 2005.

Johnston, Ian. *Beardmore Built: The Rise and Fall of a Clydeside Shipyard*. Clydebank, 1993.

_____. *Ships for a Nation, 1847-1971: John Brown and Company, Clydebank*. Dumbarton, 2000.

_____ and Johnman, Lewis. *Down the River*. Glendaruel, 2001.

Levy, Catriona. *Ardrossan Shipyards: Struggle for Survival, 1825-1983*. Ardrossan, 1983.

Lyon, David J. (comp.). *The Denny List*. 4 vols. Greenwich, 1975-1976.

Middlemiss, Norman L. *British Shipbuilding Yards. Vol. 2: Clydeside*. South Shields, 1995.

McGill, Jack. *Crisis on the Clyde, The Story of Upper Clyde Shipbuilders*. London, 1973.

O'Hara, George C. *Ironfighters, Outfitters and Bowler Hatters*. Prestwick, 1997.

Parker, George H. *At the Sharp End! A Shipbuilding Autobiography*. Glasgow, 1992.

Paulden, Sydney M. and Hawkins, Bill. *Whatever Happened at Fairfields?* London, 1969.

Peebles, Hugh B. *Warshipbuilding on the Clyde: Naval Orders and the Prosperity of the Clyde Shipbuilding Industry, 1889-1939*. Edinburgh, 1987; New ed., Edinburgh, 2000.

Reid, James M. *James Lithgow, Master of Work*. London, 1964.

Scotts of Greenock. *Two Hundred and Fifty Years of Shipbuilding by the Scotts at Greenock*. Greenock, 1961.

Shields, John. *Clyde Built: A History of Shipbuilding on the River Clyde*. Glasgow, 1943.

Slaven, Anthony. *The Development of the West of Scotland, 1750-1960*. London, 1975; New ed., London, 2009.

Thompson, Willie and Hart, Finlay. *The UCS Work-In*. Glasgow, 1972.

Walker, Fred. *Song of the Clyde: A History of Clyde Shipbuilding*. Cambridge, 1984.

Articles: Clyde and Scotland

Buxton, Neil K. "The Scottish Shipbuilding Industry between the Wars: A Comparative Study." *Business History*, X, No. 2 (1968), 101-120.

Campbell, Roy H. "Scottish Shipbuilding: Its Rise and Progress." *Scottish Geographical Magazine*, LXXX, No. 2 (1964), 107-113.

Johnman, Lewis and Murphy, Hugh. "No Light at the End of the Dock: The Long Rise and Short Life of the Firth of Clyde Dry Dock Company." *International Journal of Maritime History*, X, No. 2 (1998), 91-123.

______ and ______. "A Triumph of Failure! British Shipbuilding and the Offshore Structures Market, 1960-1990: A Case Study of Scott Lithgow Limited." *International Journal of Maritime History*, XIV, No. 1 (2002), 63-92.

______ and ______. "The Economic and Social Effects of the Interwar Depression on Clydeside Shipbuilding Communities." *International Journal of Maritime History*, XIX, No. 1 (2006), 227-254.

Johnston, Ian and Murphy, Hugh. "The Newshot Isle Project." *Scottish Geographical Journal*, CXVII, No. 3 (2001), 207-217.

McKinstry, Sam. "Transforming John Brown's Shipyard: The Drilling Rig and Offshore Fabrication Business of Marathon and UIE, 1972-1977." *Scottish Economic and Social History*, XVIII (1998), 33-60.

Murphy, Hugh. "Scott's of Greenock and Naval Procurement, 1960-77." *Mariner's Mirror*, LXXXVIII, No. 3 (2001), 196-211.

Robertson, D.J. "Labour Turnover in the Clyde Shipbuilding Industry." *Scottish Journal of Political Economy*, I, No. 1 (1954), 9-32.

Robertson, Paul L. "Shipping and Shipbuilding: The Case of William Denny and Brothers." *Business History*, VI, No. 1 (1964), 36-47.

Slaven, Anthony. "A Shipyard in Depression: John Brown's of Clydebank, 1919-1938," *Business History*, XIX, No. 2 (1977), 192-217.

Chapters in Books

Johnman, Lewis. "Scottish Shipbuilding." In Coull, James R.; Fenton, Alexander; and Veitch, Kenneth (eds.). *Boats, Fishing and the Sea. Scottish Life and Society: A Compendium of Scottish Ethnology, Vol. 4*. Edinburgh, 2008, pp. 523-544.

Slaven, Anthony. "Management and Shipbuilding. 1880-1938: Strategy and Structure in the Shipbuilding Firm on the Clyde." In Slaven, Anthony and Aldcroft, Derek H. (eds.). *Business, Banking, and Urban History: Essays in Honour of S.G. Checkland*. Edinburgh, 1982, pp. 35-53.

______. "Scottish Shipbuilders and Marine Engineers: The Evidence of Business Biography, 1860-1960." In Smout, T.C. (ed.). *Scotland and the Sea*, Edinburgh, 1992, pp. 182-191.

Books: North East Coast

Clark, Andrew. *Sunderland Shipyards*. Seaham, 1998.

Clarke, J.F. *Building Ships on the North East Coast: A Labour of Love, Risk and Pain*. 2 vols., Whitley Bay, 1997.

______. *Power on Land and Sea, 160 Years of Industrial Enterprise on Tyneside: A History of R. and W. Hawthorn Leslie and Co. Ltd., Engineers and Shipbuilders*. Newcastle, 1979.

Dougan, David. *The Shipwrights: The History of the Shipconstructors' and Shipwrights' Association, 1882-1963*. Newcastle upon Tyne, 1975.

______. *The History of North East Shipbuilding*. London, 1968.

Lingwood, J. *SD-14-The Great British Shipbuilding Success Story*. Kendal, 1976; New ed., Preston, 2004.

MacDonald, Ian and Tabner, Len. *Smith's Dock: Shipbuilders, 1908-1987.* Easington, 1986.

Middlemiss, Norman L. *British Shipbuilding Yards, Vol.1: North East Coast.* Newcastle-upon Tyne, 1993.

Pickard, Tom. *We Make Ships* [Austin and Pickersgill]. London, 1989.

Readhams, John. *Shipbuilding, 1865-1965: John Redhead, South Shields, Company History*. South Shields, 1965.

Roberts, I.P. *A Question of Construction: Capital and Labour in Wearside Shipbuilding since the 1930s*. Durham, 1998.

Smith, J.W. and Holden, T.S. *Where Ships are Born, Sunderland 1346-1946: A History of Shipbuilding on the River Wear*. Sunderland, 1946.

Stone, Ian. *Shipbuilding on Wearside*. Sunderland, 1988.

Warren, Kenneth. *Armstrongs of Elswick: Growth in Engineering and Armaments to the Merger with Vickers*. London, 1988.

Books: Belfast

Hume, John R. and Moss, Michael S. *Shipbuilders to the World: 125 Years of Harland and Wolff at Belfast, 1861-1986*. Belfast, 1986.

Middlemiss, Norman L. *British Shipbuilding Yards, Vol. 3: Belfast, Merseyside, Barrow and All Other Areas*. South Shields, 1995.

Workman, Clark. *Shipbuilding at Belfast, 1880-1933: History of Workman Clark (1928) Ltd*. Belfast, 1934.

Books: Mersey and Barrow

Cammell Laird. *Builders of Great Ships: The History of Cammell Laird and Company Ltd*. Birkenhead, 1959.

Hollett, David. *Men of Iron: The Story of Cammell Laird Shipbuilders, 1828-1991*. Birkenhead, 1992.

Roberts, David. *Life at Lairds: Memories of Working Shipyard Men*. Bebington, 1992.

Scott, John D. *Vickers: A History*, London. 1962.

Shore, Leslie M. *Vickers' Master Shipbuilder: Sir Leonard Redshaw*. Lydney, 2011.

Trebilcock, Clive. *The Vickers Brothers: Armaments and Enterprise, 1854-1914*. London, 1977.

Warren, Kenneth. *Steel, Ships and Men: Cammell Laird, 1824-1993*. Liverpool, 1998.

Books: Humber

Credland, Arthur G. *Earles of Hull, 1853-1932: Iron and Steel Shipbuilding on the Humber*. Hull, 1982.

Taylor, Mike. *Richard Dunston Limited of Thorne and Hesle: A Pictorial History*. Barnsley, 2009.

Thompson, Michael, *et al. Cook, Welton and Gemmell, Shipbuilders of Hull and Beverley, 1883-1963*, Beverley, 1999.

Books: Thames

Arnold, Anthony J. *Iron Shipbuilding on the Thames, 1832-1915: An Economic and Business History*. Aldershot, 2000.

Books: South Coast

Barnaby, K.C. *100 Years of Specialised Shipbuilding and Engineering: John I. Thornycroft Centenary, 1964*. London, 1964.

White, J. Samuel and Company. *Shipbuilding: From Smack to Frigate and from Cutter to Destroyer*. Cowes, 1928.

Williams, David L. *Maritime Heritage: Whites of Cowes, "White's Built Well-Built."* Peterborough, 1993.

Books: Devon

Harris, Len. *A Two Hundred Year History of Appledore Shipyards*. Combe Martin, 1992.

Books: East Anglia

Goodey, Charles. *The First Hundred Years: The Story of Richards Shipbuilders*. Ipswich, 1976.

Books: United Kingdom

Atkinson, Robert. *The Development and Decline of British Shipbuilding*. London, 1999.

Blake, George. *British Ships and Shipbuilders*. London, 1946.

Booz-Allen and Hamilton International BV. *British Shipbuilding 1972*. London, 1973.

Brown, David K. *A Century of Naval Construction: The History of the Royal Corps of Naval Constructors*. London, 1983.

Burton, Anthony. *The Rise and Fall of British Shipbuilding*. London, 1994.

Davies, Michael. *Belief in the Sea: State Encouragement of British Merchant Shipping and Shipbuilding: Historical Development and Future Implications*. London, 1992.

Great Britain. Parliament. *Report of the Shipbuilding Inquiry Committee, 1965-1966*. Cmd. 2937. London, 1966 (Geddes Report).

Harrison, Richard T. *Industrial Organization and Changing Technology in United Kingdom Shipbuilding*. Aldershot, 1990.

Hogwood, Brian W. *Government and Shipbuilding: The Politics of Industrial Change*. Farnborough, 1979.

Jamieson, Alan G. *Ebb Tide in the British Maritime Industries: Change and Adaptation, 1918-1990*. Exeter, 2003.

Johnman, Lewis and Murphy, Hugh. *British Shipbuilding and the State since 1918: A Political Economy of Decline*. Exeter, 2002.

Jones, Leslie. *Shipbuilding in Britain: Mainly between the Two World Wars*. Cardiff, 1957.

Lorenz, Edward H. *Economic Decline in Britain: The Shipbuilding Industry, 1890-1970*. Oxford, 1991.

Matsumoto, Miwao. *Technology Gatekeepers for War and Peace: The British Ship Revolution and Japanese Industrialization*. Basingstoke, 2006.

Parkinson, John R. *The Economics of Shipbuilding in the United Kingdom*. Cambridge, 1960.

Pollard, Sidney and Robertson, Paul L. *The British Shipbuilding Industry, 1890-1914*. Cambridge, MA, 1979.

Pugh, Philip. *The Cost of Seapower: The Influence of Money on Naval Affairs from 1815 to the Present Day*. London, 1986.

Ritchie, L.A. (ed.). *The Shipbuilding Industry: A Guide to Historical Records*. Manchester, 1992.

Slaven, Anthony. *British Shipbuilding, 1500-2010: A History*. Lancaster, 2013.

Speed, Keith. *Sea Change: The Battle for the Falklands and the Future of Britain's Navy*. Bath, 1982.

Starkey, David J. and Murphy, Hugh (eds.). *Beyond Shipbuilding and Shipping: Britain's Ancillary Maritime Interests in the Twentieth Century*. Hull, 2007.

Articles: United Kingdom

Arnold, A.J. "'Of Not Much Significance:' The Inclusion of Naval and Foreign Business in the UK Shipbuilding Output Series, 1825-1914." *International Journal of Maritime History*, XI, No. 2 (1999), 61-86.

Johnman, Lewis. "The Privatisation of British Shipbuilders." *International Journal of Maritime History*. VIII, No. 2 (1996), 1-31.

_____ and Murphy, Hugh. "The Norwegian Market for British Shipbuilding, 1945-1967." *Scandinavian Economic History Review*, XLVI, No. 2 (1998), 55-78.

_____ and _________. "A Very British Institution! A Study in Under-Capitalisation: The Role of the Ship Mortgage Finance Company in Post-Delivery Credit Financing within British Shipbuilding, 1951-67." *Financial History Review*, VI, No. 2 (1999), 203-221.

_____ and _____. "The Rationalisation of Warship Building in the United Kingdom, 1945-2000." *Journal of Strategic Studies*, XXIV, No. 3 (2001), 107-127.

_____ and _____. "The British Merchant Mission in the United States and British Merchant Shipbuilding during the Second World War." *The Northern Mariner/Le Marin du Nord*, XII, No. 3 (2002), 1-15.

_____ and _____. "Subsidy and Treasury: The Trade Facilities Act and the UK Shipbuilding Industry in the 1920s." *Contemporary British History*, XXII, No. 1 (2008), 89-110.

McGoldrick, James. "Industrial Relations and the Division of Labour in the Shipbuilding Industry since the War." *British Journal of Industrial Relations*, XXI, No. 2 (1983), 197-220.

Murphy, Hugh. "'From the Crinoline to the Boilersuit:' Women Workers in British Shipbuilding during the Second World War." *Contemporary British History*, XIII, No. 4 (1999), 82-104.

_____. "The Health of Electric Arc Welders and the Adoption of Arc Welding in the British Shipbuilding Industry, 1930-1951." *International Journal of Maritime History*, XVII, No. 1 (2005), 69-90.

_____. "Charting an Industry's Rise and Fall." In *Tales from the Shipyard: Britain's Shipbuilding Heritage on Film*. British Film Institute, London, 2011.

_____. "The British Shipbuilding Industry during the Great War: A Contextual Overview Incorporating Standardization and the National Shipyards, 1916-1920." *International Journal of Maritime History*, XXIV, No. 2 (2012), 19-68.

Parkinson, John R. "The Financial Prospects of Shipbuilding after Geddes." *Journal of Industrial Economics*, XVII, No. 1 (1968), 1-17.

_____. "Trends in the Output and Export of Merchant Ships." *Scottish Journal of Political Economy*, III, No. 3 (1956), 234-245.

Slaven, Anthony. "British Shipbuilding: Market Trends and Order Book Patterns between the Wars." *Journal of Transport History*, 3rd ser., III, No. 2 (1982), 37-61.

Stopford, R.M. and Barton, J.R. "Economic Problems of Shipbuilding and the State." *Maritime Policy and Management*, XIII, No. 1 (1986), 27-44.

Todd, Daniel. "Regional Variations in Naval Construction: The British Experience, 1895-1996." *Regional Studies*, XV, No. 2 (1981), 123-142.

Chapters in Books

Johnman, Lewis. "The Shipbuilding Industry." In Mercer, Helen; Rollins, Neil; and Tomlinson, Jim (eds.). *Labour Governments and Private Industry: The Experience of 1945-1951*. Edinburgh, 1992, pp. 186-211.

______. "Old Attitudes and New Technology: British Shipbuilding, 1945-1965." In Van Royen, Paul C.; Fischer, Lewis R.; and Williams, David M. (eds.). *Frutta di Mare: Evolution and Revolution in the Maritime World in the 19th and 20th Centuries*. Amsterdam, 1998, pp. 133-152.

______. "Public Intervention and the Hollowing out of the British Shipbuilding Industry: The Swan Hunter Closure." In Erstsvåg, Randi; Starkey, David J.; and Austbø, Anne Tove. (eds.). *Maritime Industries and Public Intervention*. Stavanger, 2002, pp. 223-239.

______ and Murphy, Hugh. "Welding and the British Shipbuilding Industry: A Major Constraint?" In Harding, Richard (ed.). *The Royal Navy, 1930-2000: Innovation and Defence*. London, 2005, pp. 89-116.

______ and ______. "The Rationalisation of Warship Building in the United Kingdom, 1945-2000." *Journal of Strategic Studies*, XXIV, No. 3 (2001), 107-127. Reprinted in Lambert, Andrew D. (ed.). *Naval History, 1850-Present, Vol. II*. Aldershot, 2007.

______ and ______. "The Development of the British Ship Repair Industry, 1945-1985: An Overview." In Starkey, David J. and Murphy, Hugh (eds.). *Beyond Shipbuilding and Shipping: Britain's Ancillary Maritime Industries in the Twentieth Century*. Hull, 2007, pp. 81-106.

______ and ______. "The Rationalisation of Slow Speed Marine Diesel Engine Building in the UK, 1912-1990." In Starkey, David J. and Murphy, Hugh (eds.). *Beyond Shipbuilding and Shipping: Britain's Ancillary Maritime Interests in the Twentieth Century*. Hull, 2007, pp. 29-64.

Lorenz, Edward H. and Wilkinson, Frank. "Shipbuilding and British Economic Decline, 1880-1965." In Elbaum, Bernard and Lazonick, William (eds.). *The Decline of the British Economy*. Oxford, 1986, pp. 109-134.

McMurray, H. Campbell. "The Records of the Shipbuilders and Repairers National Association." *Business Archives*, XLV (1975), 18-23.

Murphy, Hugh; Johnman, Lewis; and Ritchie, L.A. "De-accessioning and British Shipbuilding Records." In Richmond, Lesley M. (ed.). *"Pioneering New Frontiers:" An International Exploration of Current Initiatives in Business Archives*. London, 1997, pp. 159-168.

Slaven, Anthony. "Shipbuilding Industry Organisations and Policies, 1920-1977." In Slaven, Anthony and Kuuse, Jan (eds.). *Scottish and Scandinavian Shipbuilding: Development Problems in Historical Perspective*. Gothenburg, 1981.

______. "Self-Liquidation: The National Shipbuilders Security Ltd. and British Shipbuilding in the 1930s." In Palmer, Sarah and Williams, Glyndwr (eds.). *Charted and Uncharted Waters: Proceedings of a Conference on British Maritime History*. London, 1982, pp. 125-147.

______. "Marketing Opportunities and Marketing Practices: The Eclipse of British Shipbuilding, 1957-1976." In Fischer, Lewis R. (ed.). *From Wheel House to Counting House: Essays in Maritime Business History in Honour of Professor Peter Neville Davies*. St. John's, 1992, pp. 125-151.

______. "Modern British Shipbuilding, 1800-1990." In Ritchie, L.A. (ed.). *The Shipbuilding Industry: A Guide to Historical Records*. Manchester, 1992, pp. 1-24.

______. "From Rationalisation to Nationalisation: The Capacity Problem and Strategies for Survival in British Shipbuilding, 1920-1977." In Feldenkirchen, Wilfried; Schönert-Röhlk, Frauke; and Schulz, Günther (eds.). *Hans Pohl Liber Amicorum; Wirtschaft Gesellschaft Unternehmen*. Stuttgart, 1995, pp. 1128-1155.

Thomas, Dennis. "Shipbuilding: Demand Linkage and Industrial Decline." In Williams, Karel; Williams, John L.; and Thomas, Dennis (eds.). *Why are the British Bad at Manufacturing?* London, 1983, pp. 179-216.

Printed and bound by CPI Group (UK) Ltd, Croydon, CR0 4YY
13/09/2026
14961099-0002